Y0-DXF-205
3 0000 002 605 388
LIBRARY OF MICHIGAN

The Social Costs of Genetic Welfare

The Social Costs of Genetic Welfare

Marque-Luisa Miringoff

RUTGERS UNIVERSITY PRESS

New Brunswick, New Jersey

Manufactured in the United States of America

LIBRARY OF CONGRESS CATALOGING-IN-PUBLICATION DATA

Miringoff, Marque-Luisa, 1947–
The social costs of genetic welfare / Marque-Luisa Miringoff.
p. cm.
Includes bibliographical references and index.
ISBN 0-8135-1706-0 (cloth) — ISBN 0-8135-1707-9 (pbk.)
1. Genetic engineering—Moral and ethical aspects. 2. Human reproductive technology—Moral and ethical aspects. 3. Handicapped—Public opinion. I. Title.
RB155.M56 1991
174'.25—dc20 *91-16793*
CIP

British Cataloging-in-Publication information available

To Helen and Hy
who taught me

Contents

Acknowledgments

The Chinese have a saying that goes "May you live in interesting times." Whether this is a blessing or a curse, I am not sure, but clearly we live in "interesting times." The changes that have challenged this society are incalculable. Some of us have felt the urge to record and comment on these interesting times. This is not a process that happens alone. It is a shared event, and I have been fortunate to have had the support and encouragement of many.

My gentle and kind friend, Lilo Stern, associate professor of anthropology at Vassar College, read this manuscript more times than I care to admit or record. At each point, from my earliest drafts to the final form, she offered me her wisdom, her judgment, and her very pointed criticism. Her friendship and support have been infinite. I thank her with all my heart, for both.

My colleague and good friend in the Sociology Department at Vassar College, Professor Eileen Leonard, offered me similar, seemingly limitless, patience and encouragement. "How's the book going?" she would ask, at tactfully spaced intervals, giving me a thoughtful push in the direction of completion. She too read several drafts of the manuscript and helped immeasurably to sharpen and focus its direction.

My friend of long years, Deborah Stone, an educator in the truest sense of the word, also gave me great encouragement. Her sensitive reading of the manuscript helped me to fine-tune many details. To a friend of newer years, Jean Stewart, I thank for her several visits to my classes. Her own writings on the subject of disability, her hard-nosed activism in the disability rights movement, and her dedication served as a model, strengthening my conviction in the urgency of my subject.

I owe a very great intellectual debt to the Hastings Center, a research institute for the study of bioethical issues. The center gave me the opportunity to serve as a research Fellow, to use its remarkable resources, and to consult with its extraordinary and dedicated faculty. In particular, I would like to thank Daniel Callahan, director of the Hastings Center, for reading the manuscript in one of its later drafts, and for steering me in several important directions.

Chapter 3 is a revised version of "Genetic Intervention and the Problem of Stigma" *Policy Studies Review* 8, no. 2 (Winter 1989).

Vassar College was extremely helpful in providing the support and facilities needed for the completion of this work. I would like to thank the reference staff of the library, and Shirley Maul, in particular, for assisting in the retrieval of countless materials. All of the staff were extraordinarily gracious and helpful.

The Vassar College administration provided generous leave time and support. I would particularly like to thank former president Virginia Smith, who gave me the opportunity to pursue this book in its early stages, and current president Frances Fergusson, and Nancy Dye, dean of the faculty, for providing the leave time to complete this project. In addition, I owe a debt of gratitude to my Vassar College students, particularly those who shared in my explorations of disability, medicine, and social welfare. Their idealism and sense of justice have always inspired me.

Ruth Hubbard, professor of biology at Harvard University, read one of the final drafts of the manuscript; I owe her my sincerest thanks and respect. To have detailed her comments and recommendations, as she did, with such precision and care, left me in awe. She sets a true standard, difficult to match, for academic thoughtfulness and clarity of communication.

To Marlie Wasserman, my editor, I owe an enormous debt. For "hanging in there" with me, over so many years, she deserves recognition for actions beyond the call of duty. I would also like to thank her for her kindness and consideration, and for her serious and thoughtful reflections on my work.

My final and greatest debt is to my husband, Marc L. Miringoff, who is the assistant dean of Fordham University's Graduate School of Social Services and director of the Fordham Institute for Innovation in Social Policy. There has been no day in our life together that he has not strengthened me. His presence is on each page of this book as surely as his presence is in every part of my life. To him I give my deepest love and gratitude.

Introduction

Several years ago, I developed a course on the subject of physical disability. During my preparation, it became increasingly clear that social attitudes toward people with disabilities were crystallizing into two very different points of view. The first, a progressive development from my perspective, was emerging out of the civil rights movement of disability advocates. This approach stressed improved access to buildings and facilities, mainstreaming strategies, and efforts to increase both the visibility and the social opportunities of disabled people. The potential contributions of people with disabilities and the need to break down the social barriers that might prevent these contributions were central issues.

The second and more recent stream of thought that I uncovered emphasized not the social situation of disabled individuals, but rather the possibilities for preventing their future existence through the technologies of genetic and reproductive engineering. This view argued that through the application of scientific knowledge we could minimize the number of disabled people who would ever enter society. Their lives were portrayed as extraordinary tragedies, unacceptable and potentially avoidable lapses in human perfectability. Thus, rather than seeking to enhance the living conditions of people with disabilities, as the first view proposed, this genetic perspective communicated an intensified sense of horror at the very fact of disability, and a new sense of urgency to utilize and develop our technology in order to prevent its appearance.

These two views, one a social vision seeking to reconstitute the environment in order to accommodate the special needs of social groups, the other a genetic vision seeking to excise or biologically refashion the problem, are fundamentally at odds. While it is clearly possible to pursue genetic approaches at the same time that one extends civil liberties, the mind-sets underlying the two visions are nonetheless in conflict. The social view seeks to expand the range of accepted human form and behavior, whereas the genetic vision seeks to ensure biological fitness. The social view seeks to maximize the talents and contributions of people with disabilities; the genetic vision dwells on losses in functioning and mobility. The social view seeks to integrate those with disabilities into the mainstream of society; the genetic vision seeks to prevent the very existence of such individuals.

These two contrasting views serve to illustrate the central concern of this book, namely, that a distinctive worldview is evolving with our new genetic and reproductive technological capacities. This emerging genetic worldview challenges more socially oriented views and may have serious implications for selected social groups.

I have entitled the genetic worldview *Genetic Welfare*. This perspective envisions the public good as best served by genetic or biological intervention. Its focus is on our genetic inheritance and our ability to enhance that inheritance through technological intervention. I have chosen the term *Genetic Welfare* because it serves as a contrast to what is traditionally called *Social Welfare*, an orientation that seeks to modify the social environment, its laws, policies, institutions, its fairness, and equity. In a Social Welfare perspective, it is the social structure that is the central focus of concern.

The worldview of Genetic Welfare is evolving with the acceleration of new genetic and reproductive technologies. In vitro fertilization, genetic screening, surrogacy, embryo transfer, and sex selection are becoming routinely available reproductive technologies. Recombinant DNA and genome mapping are expanding the available methods of diagnosis and analysis.[1] Such techniques are increasingly part of the everyday landscape of our lives, familiar signposts of our scientific progress and prowess.

Because of their technical allure and extraordinary rate of progress, genetic strategies are rapidly becoming more significant than socially oriented strategies. Genetic and reproductive technologies are increasingly the methods of choice; they appear simpler, cleaner, neater, and more accurate. They are *scientific,* with all that that term connotes. Social Welfare strategies, alternatively, now seem old-fashioned, less precise, and less focused; they are being incrementally edged out by new technological developments.

In this book I explore the emergence of Genetic Welfare, its tendency to displace Social Welfare strategies, and the implications of this significant social change. In Part I, I examine the rise of Genetic Welfare and its increasing acceptance as a vision of the world. In Part II, I consider the social problems that may result from this shift in worldview, specifically issues of stigma, powerlessness, and alienation. Finally, in Part III, I assess the scientific and citizen responses to genetic and reproductive technologies. The first efforts to regulate these technologies and the first citizen protests are explored.

The orientation of this volume is sociological; more specifically, I take a social-problems and public-policy perspective. Thus this work is deliberately *problem seeking*. I focus on the potential hardships that Genetic Welfare may entail for social groups, not on its potential benefits. The contributions of genetic and reproductive developments are well known and have been documented in countless books and articles. As a society, we follow genetic and reproductive discoveries with an avid eye, eager to observe what new "miracle" science will create for us. This book acknowledges the importance of these scientific developments and their potential contributions to our physical well-being, but I leave the elaboration of their benefits to other authors. My focus is on the underside of the biological revolution, its latent effects on social groups, its impact on our vision of the world, and its influence on our acceptance and tolerance of those who are outside the mainstream.[2]

An important thread running through the book is the question of balance. If genetic visions become increasingly more important than social perspectives, who will be harmed? My essential concern is that Genetic Welfare ultimately may serve to undermine the status of disadvantaged groups. A genetic worldview has the potential to newly stigmatize individuals with disabilities, newly disempower disadvantaged groups, and newly alienate medical relationships.

In the arena of stigma and disability, a fragile process has recently been achieved through civil rights legislation and the work of advocacy groups. Mainstreaming has brought to public consciousness a new awareness of the skills and contributions of those with disabilities. Yet if genetic approaches come to predominate, a new aversion to disability may arise, as interventive methods for identification and prevention become commonplace. Our desire to "optimize" human beings puts new pressures on families and physicians. New modes of stigmatizing disability emerge through genetic labels, and our tolerance for the existence of human difference is whittled away.

In similar fashion, civil rights activists of the 1960s and early 1970s sought to empower other disadvantaged groups typically outside the mainstream—the poor, minorities, and women. Although such political progress was negligible in the 1980s, some accomplishments were maintained. Yet now a new mode of disenfranchisement is emerging through the development of genetic and reproductive technologies. These new technologies have an elite bent, primarily serving the well-to-do and ignoring the poor. Moreover, in seeking to "perfect" society, the characteristics of the powerful typically are targeted for reproduction, the "less-than-perfect" bypassed; while sex selection for males, in preference to females, may become yet another mode of disempowerment.

Finally, the medical community has been struggling in the last several decades to overcome an increasing sense of alienation on the part of patients. Developments such as "informed consent," "living wills," and "hospital ethics committees" have sought to newly enfranchise the patient, enhancing patient involvement and decision making. Yet genetic and reproductive technologies are now taking medicine in an opposite direction. They introduce yet another form of high technology, placing a mountain of machinery between doctor and patient. The patient increasingly becomes an object for genetic investigation and experimentation. The goals of humanity and dignity in medicine are further challenged.

How we, as a society, will address these problems is a critical item for our future public-policy agenda. Thus far, efforts to regulate genetic and reproductive investigations have been slight. The Congress and the Executive Branch have been reluctant to hinder the work of physicians and scientists. Physicians and scientists, in turn, wish to maintain their freedom of inquiry. Citizen groups have protested these new developments with varying degrees of success. Some have sought to halt research in their communities, while still other groups have attempted to define public priorities. All these efforts are in a preliminary phase.

We are still on the threshold of assessing these various new technologies, their multiple social meanings, and their implications for public policy. Still needed is a holistic perspective, a way to understand the worldview that accompanies genetic intervention, the worldview it will displace, and the social problems that will ensue. In our conceptualization of our problems and our visualization of their solutions, we are now forging a dynamic conflict between genetic and social perspectives. The balance that emerges will have serious consequences for the nature of the society we create in the twenty-first century.

Genetic and reproductive technologies have the capacity to alter the very fabric of our social structure, fundamentally shifting our worldview and transforming our social behavior. These techniques represent new tools which may be used to modify both ourselves and our environment. Our problems may be solved in new ways. A new logic is generated; a new calculus of costs and benefits is created. It is the task of this book to explore this new calculus, to assess its impact on social groups, its changing manifestations over time, and its implications for public policy. In so doing, I hope to suggest what new social problems we may face in the future and what new forms of public policy and sensitivity we will need in order to address these problems.

PART I

Defining Genetic Welfare

1

An Emerging Worldview

God: *Oh, I remember well: you're Job, my Patient.*
How are you now? I trust you're quite recovered,
And feel no ill effects from what I gave you.

Job: *Gave me in truth: I like the frank admission.*
I am a name for being put upon.
But, yes, I'm fine, except for now and then
A reminiscent twinge of rheumatism.
The let-up's heavenly. You perhaps will tell us
If that is all there is to be of Heaven,
Escape from so great pains of life on earth
It gives a sense of let-up calculated
To last a fellow an Eternity.

—ROBERT FROST,
A Masque of Reason

In our ancient past, illness demanded both resignation and courage. As in the Book of Job, patients were advised to have fortitude in the presence of unjust pain and suffering. With the development of modern medicine, however, this vision changed. The control of disease replaced the notion of endurance.

The worldview of modern medicine, as it evolved over the past century, embodied the belief that, ultimately, virtually every disease would yield to the rationality of science. Germ theory, the understanding that microorganisms cause disease, and specific etiology, the search for a single and identifiable cause for each illness, served as the foundations for an effective new paradigm of medical diagnosis and treatment.

Public health measures similarly altered our perception of illness. Since the nineteenth century, advances in sanitation, water supply, nutrition, housing, and working conditions functioned as powerful new tools in the reduction of suffering and mortality.[1] The prevention of disease came within our purview.

One by one, infectious diseases were brought under control. Epidemics of diphtheria, tuberculosis, polio, and other diseases were diminished or eliminated. Great scientists and physicians became part of our history, as important as Columbus or Caesar. The ability to govern fate altered our perception of ourselves—from victims to heroes.

Throughout this history, however, certain illnesses defied these medical advances. Genetic diseases and reproductive disorders, in particular, were relatively resistant to medical intervention. Now, these too are yielding. A variety of new approaches are at hand, and once again we have the opportunity to refashion our vision of the world.

An array of new interventions to prevent or treat genetic and reproductive disorders is emerging; refinements are appearing at an accelerating pace. These developments may be applied at many points in the life cycle. Prior to the conception of a child, for example, genetic screening of potential parents can project the probability and sometimes the certainty of bearing a child with a genetic disability. Childbearing decisions can be informed on this basis.

After conception of a child, genetic screening procedures can discern the existence of several hundred genetic disorders.[2] If a disease is identified, abortion becomes an option. Ultimately, more sophisticated technologies, now in development, such as gene therapy and gene surgery may permit the alteration of genes in utero, thereby preventing the appearance of a disease entirely.

Following birth, several forms of neonatal screening can detect specific genetic conditions. In certain cases, as with the disease phenylketonuria (PKU), treatments can be prescribed. Diets and drugs may be used to prevent the onset of symptoms and degenerative effects.

Other new measures address reproductive incapacities. Current technologies include artificial insemination, in vitro fertilization, and embryo transfer, all of which may be used by natural or surrogate parents. More futuristic technologies may include cloning, that is, the duplication of genetic material, and ectogenesis, the complete gestation of a fetus outside the womb. By screening both parents and embryos, these reproductive technologies, too, could be used to prevent genetic disorders.

Finally, recombinant DNA, the process of gene splicing, also offers the possibility of genetic intervention. This new technology has led to the creation of new organisms and treatments. Medical, industrial, and agricultural uses are expanding as these techniques are being rapidly refined. More recently, recombinant DNA investigations have also initiated the process of genetic "mapping," permitting the identification of genetic markers "that allow scientists to track the inheritance of disease from one generation to the next."[3]

Taken together, these many new technologies fundamentally alter our medical and scientific abilities. They permit interventions where no treatments previously existed. They permit choices where endurance and the acceptance of fate were once required. With the ability to "detect," "identify," "create," "map," or "alter" human beings, we are presented with decisions prior generations did not face. We now have the capacity, in a sense, to select among human beings and their characteristics, to produce human beings where the possibility did not previously exist.

Logically, the introduction of so many new choices must impinge on our vision of the world. With the ability to control dilemmas once beyond our governance, we must eventually come to see ourselves and others from a somewhat different perspective. Just as the elimination of disease in the early part of this century fundamentally shifted our perceptions of life expectancy and life quality, so too will modern technologies likely influence our interpretation of our capacities and abilities. Such changes must ultimately reflect upon considerations of our general well-being or what is our greatest good. These deliberations form the basis of an altered understanding of our social universe. Thus, it will be argued, a distinctive worldview is emerging from our new technological abilities in genetic and reproductive engineering; this worldview is what will be called Genetic Welfare.

GENETIC WELFARE: A WORLDVIEW

As defined in this book, Genetic Welfare is the desire to improve the human condition through genetic and reproductive intervention. Its proponents are those who seek to modify the genetic foundation of human life, to bring about a healthier society, with fewer human problems.

The term Genetic Welfare was chosen for several reasons. First, the word *welfare* is intended to connote considerations of action and intervention as well as human betterment. Societies *provide* for their population's welfare through legislation and public policy. Additionally, Genetic Welfare was selected to serve as a contrast to what is traditionally termed Social Welfare. Social Welfare, historically, has sought to improve human life by alterations in the social environment, through organizational and institutional change. In a Social Welfare perspective, problems are perceived as embedded in society; solutions are typically located in the social structure. In Genetic Welfare, by contrast, change is sought at the genetic level; its purpose, to improve the human species.

In this definition, Genetic Welfare reflects a relative emphasis. It is not so much the total dominance of genetic considerations that is at issue, but their *rising* significance. The argument will be made that in this emerging worldview, genetic assessments and their interventions are given increasing importance while social analyses and actions are receding in consideration. In an array of issues involving birth, death, disease, disability, and the quality of life, we now may have to choose between genetic and social iterventions. This choice is tilting toward the genetic, reflecting a shift in our vision of our abilities.

It will be argued further that this genetic worldview is evolving incrementally. As each new technology advances, our abilities expand. Our increased understanding of our many new tools gives rise to a new sense of empowerment, a new order of priorities, and a new perception of the future. These visions ultimately may constitute a fundamental change from the past; one which is distinctly different from the "old" social welfare.

GENETIC WELFARE VERSUS SOCIAL WELFARE

In the twentieth century, social problems traditionally have been handled through broad social welfare policies. Income maintenance programs, employment strategies, housing subsidies, and affirmative action are examples. These policies assume that human behavior can be altered through social change. With the emergence of Genetic Welfare, new policies may come to take their place.

Today, the techniques of genetic and reproductive engineering are often viewed as more practical alternatives to the cumbersome processes of social change. Verle Headings, for example, has argued that "disillusionment with social engineering as a means of modifying self-defeating forms of behavior prompts some scientists to dream of genetic engineering as an alternative means to this goal."[4] As Headings notes, genetic techniques now promise more successful resolutions to our increasingly complex problems.

Several years ago, Alvin Weinberg advanced the idea of a quick "technological fix" to our social problems, suggesting that for many social issues, technological approaches would be both more effective and efficient than expensive social interventions.[5] More recently, this view has been seen as both possible and desirable through the techniques of genetic and reproductive engineering. As scientist James Danielli argues, "Social scientists pin their hopes . . . on the possibility of improving human institutions and environments. But . . . this is a most dubious proposition. . . . We must consider other possibilities if civilization is to advance. . . . And these other possibilities lie in genetic engineering if they lie anywhere in the world."[6] To some scientists, genetic engineering represents our final chance to solve our social problems.

New genetic interventions are alluring for many reasons. They appear to reduce the unwieldy nature of social issues to the relative clarity of genetic defect. The crosshatched complexity of class, status, income, occupation, education, age, race, sex, religion, ethnicity, and other facets of social experience are daunting. They are untidy complications which only rarely permit extensive generalization. To substitute the more measurable precision of the laboratory has its obvious appeal.

Further, our new technological abilities come at a time when there is great consensus that old models have failed. Classic New Deal and Great Society approaches are widely questioned. To many, government seems unable to address our social ills.[7] Social programs, like social variables, are sloppy, often costly, and generally resistant to methods of accountability. Genetics and the pristine measures of science are thus tempting alternatives.

Finally, unlike the "old" apparatus of social welfare, the biological revolution has the aura of futurity about it. It has the feel of progress; it suggests forward movement; it is new. With the glamour of "breakthrough" science and the prestige of Nobel Prizes, it makes headlines.

The inducements toward a path of genetic intervention are many. If they are taken together, their logical outcome may be an altered vision of our capacities. The question that arises, then, is what specific form will this worldview assume? What new precepts will be taken as "given," and what new preoccupations will come to prevail?

In the following sections, I consider several new emphases in public debate and social policy that are indicants of this shifting worldview. These comprise first, the widely acknowledged reemergence of *biological determinism,* an ideological movement that serves as an underpinning to many genetic interventions; second, the inclination toward creating *genetic rights and duties,* a new perception of our individual obligations toward ensuring genetic health; and third, a reevaluation of our social responsibilities toward those who are genetically ill, a reassessment of our *genetic burden*.

BIOLOGICAL DETERMINISM

One of the most frequently noted trends in contemporary scientific thought is the reemergence of biological determinism. The discipline of sociobiology and the continuing assessments of race and IQ are examples. Edward O. Wilson, Richard Dawkins, Arthur Jensen, Hans J. Eysenck, and Richard Herrnstein are among the leading theorists of the movement.[8] Critics such as Steven Jay Gould, Ruth Hubbard, Marian Lowe, Richard Lewontin, Steven Rose, and Leon Kamin have frequently documented the expanding dominance of this paradigm in contemporary assessments of human behavior.[9]

Problems of all varieties now call forth biological explanations and assessments. Female aggression is dubbed premenstrual; criminality is attributed to an extra Y chromosome; IQ is deemed inherited; alcoholism is a genetic proclivity; childhood hyperactivity is a genetic defect. Increasingly, our problems are defined as *bio-social,* with the "bio" of the equation viewed as causal and the "social" as mere outcome.[10]

Our problems and their solutions seem more clearly understood in biological terms. Certain people are *ill*; that is why they act as they do. Human beings are *genetically prone* to particular behaviors; this explains both our strengths and our weaknesses. Social and moral decisions may then be made on this basis. As Edward O. Wilson has advised, "the time has come for ethics to be removed temporarily from the hands of philosophers and be biologized."[11]

This rising biological determinism has significant implications for social thought and social policy. At the most fundamental level, it suggests an impending shift from a social to an individualistic orientation. In *Not in Our Genes*, for example, Lewontin, Rose, and Kamin argue that biological determinists "are committed to the view that individuals are ontologically prior to society and that the characteristics of individuals are a consequence of their biology."[12] Further, they argue, this vantage point is politically conservative. Biologism evokes, by its very precepts, the notion of individual responsibility for social issues. The impact of social structure, social inequality, and social intervention, therefore, is relatively diminished in importance.

This shift, from social to individualistic problem solving, may be viewed as a classic example of "blaming the victim." In the early 1970s, for example, William Ryan, in his book of that title, anguished over what he perceived as a rising trend in public policy that was transforming social issues into individualistic victim-blaming solutions. To Ryan, problems such as poverty, illegitimacy, and illiteracy were increasingly being seen not as outcomes of social conditions, but as faults of character, weaknesses of judgment, lack of willpower, inability to defer gratification, and individual incapacities in many forms.[13]

Ironically, Ryan considered this new form of individualistic policy-making to be a replacement for older forms of individualistic thinking such as eugenics and biological determinism. Yet today we are once again returning to what Ryan perceived as the older form of individualistic thinking. Biological victim blaming has once again entered the forefront of ideological thought.

One of the most striking contemporary examples of biological victim blaming occurred during the 1970s corporate trend toward genetic screening. At that time, companies with unsafe toxic conditions began genetically screening potential employees, with an eye to weeding out toxic-sensitive individuals. They sought, instead, to hire individuals who were genetically better able to withstand environmental pollutants.

For these companies, the social policy imperative to provide safe and nontoxic working conditions was subordinated to the less expensive and simpler solution of selecting genetically resilient individuals. Dr. Eula Bingham, former director of the Occupational and Safety Administration, observed that we had begun "to place the burden of controlling toxic substances on individuals who are denied employment because of their supposed sensitivity, rather than on companies who should be cleaning up the workplace for all."[14]

Trade unions have been particularly sensitive to the worldview fostered by such practices. Anthony Mazzocchi, former vice president in charge of health and safety in the Oil, Chemical and Atomic Workers notes, "Management benefits from genetic screening because it creates a public consciousness that says, it's not the polluted workplace that's to blame for occupational health problems, it's the makeup of the worker."[15]

After the initial outcry over these practices, investigations were conducted by Congress and the Office of Technology Assessment.[16] But with limited knowledge of the frequency of screening, official bodies could do little more than decry the practice and the controversy subsided. In the 1980s, however, new concerns over industrial genetic screening emerged. With the increasing information provided by genetic markers, companies claim they will soon have the capacity to predict both the longevity and the expected psychological and physical productivity of workers. Severely debilitating or fatal diseases such as Huntington's chorea already can be predicted, and selected populations are now being tested; studies on other serious diseases are planned.

These new practices enhance the search for corporate productivity in the genetic characteristics of workers, rather than in the conditions of the working environment. According to Gina Kolata, "with the likely development of tests to predict susceptibility to diseases such as Alzheimer's, the same questions that were raised about genetic screening in the workplace are being asked again and with more urgency than ever."[17]

Critics argue that in the process of implementing widespread genetic testing, the very nature of work is challenged; the power of employers is vastly increased, the power of employees further diminished. Mark Rothstein, for example, notes of this practice: "We are talking about conditions for which all workers are at risk. It broadens the question and, I think, points out that the use of ever more encompassing medical tests raises serious issues."[18] The insurability and employability of workers is now an issue that will have to be addressed by policymakers.

Another prominent example of biological determinism lies in the genetic assessment of behavioral disorders. David E.

Comings, for example, in his 1989 Presidential Address to the American Society of Human Genetics, contrasts genetic versus behavioral science approaches in the treatment of antisocial behavior. Comings illustrates the behavioral science approach through Ken Magid and Carole A. McKelvey's 1987 book, *High Risk Children Without a Conscience*. Magid and McKelvey argue that antisocial behavior is due to the separation of children from their parents at an early age and make a strong plea for viewing society as a source of the problem. They write: "The chances for increasing numbers of psychopaths are escalating. We must search for answers to the pressing social problems that are helping to create unattached children. We must learn how to prevent unattached children. The solutions will not be easy or cheap, but they must be found."[19] Comings challenges this socially oriented perspective with a vision founded in genetics. He argues:

> A genetic viewpoint is different and would suggest that a disinhibition-disorder gene carried by a parent could result in marital chaos and separation and that *it is this inherited gene—not the fact that the parents separated—that causes antisocial personality in the child*. . . . The difference in cost to society of the two approaches can be enormous. With the [Magid-McKelvey] approach, billions of dollars could be spent in social programs that might have no appreciable effect on the incidence of antisocial personality. In the genetic approach, probes to identify the responsible gene in a child with conduct disorder could identify a small number of high-risk individuals who could be appropriately treated at relatively low cost.[20]

To Comings, the source of the problem is clearly the individual, specifically in a flawed gene, not in the social conditions of family breakup. Indeed Comings seems to argue that a specific gene can *cause* "marital chaos." Further, he sees the genetic approach as cost-saving to the society and more focused. Family disorganization need not be addressed. Genetic intervention will solve the problem.

The search for genetic explanations and solutions to the problems of work and behavior are but two examples of the trend toward biological determinism. This trend leads logically to the attribution of cause at the individual level; individualistic problem solving is its natural consequence. Such orientations constitute the underpinnings of Genetic Welfare. They set the stage for a fundamental reconceptualization of our rights and duties.

GENETIC RIGHTS AND DUTIES

Emerging out of the new emphasis on biological determinism and individualistic problem solving is a slowly changing perception of our rights and duties. Individuals are increasingly seen as having a *right* to genetic health; parents have a *duty* to ensure this outcome.[21] The technological ability to accomplish these outcomes imposes a new imperative toward ensuring our genetic well-being.

The contemporary version of genetic rights and duties was first formulated in the 1930s by Hermann J. Muller, the Nobel Prize-winning biologist. Muller argued that parents had an obligation to behave circumspectly in the production of offspring. "What is most needed," Muller wrote, "is an extension of the feeling of social responsibility to the field of reproduction. . . . When people come to realize that in some measure their gifts as well as their failings and difficulties—physical, intellectual, temperamental—have genetic bases and that social approval or disapproval will be accorded them if they take these matters into account in deciding how much of a family to beget, a big step forward will have been taken in the motivation of human reproduction."[22]

In the 1970s, this vision was forcefully restated by scientist Bentley Glass in what is perhaps the most famous assertion of these rights and duties. Glass wrote: "In a world where each pair must be limited, on the average, to two offspring and no more, the right that must become paramount is not the right to procreate, but rather the right of each child to be born with a sound physical and mental constitution, based on a sound genotype."[23]

In Glass's vision of the future, the right to procreate was replaced with a new and more compelling right: the right to be free of genetic disability. The onus, he argued, was to be placed on parents to behave in a genetically dutiful manner. According to some, this duty is now evolving into an ethical precept. As Werner Heim asks, "Should an eleventh commandment have been promulgated: Breed not, ye who carry defects?"[24]

In the 1990s, Glass's vision appears to be assuming an increasing reality. Technologies once seen as choices are taking on an increasingly obligatory tone.[25] Through technological advances, social pressures to maintain and perfect our health are mounting. According to the President's Commission for the Study of Ethical Problems in Medicine and Biomedical and Behavioral Research, the fetus is now more often seen as "patient, rather than the inaccessible and largely unknown predecessor of an infant."[26] The outcome of this change, according to the commission, will be "more-demanding social expectations of parents in promoting the welfare of the fetus."[27] Further, it is concluded, although "developments in prenatal therapy increase the range of technically feasible options, social pressures may severely limit parents' freedom to refrain from choosing certain options."[28]

The desire to enforce genetically responsible behavior has taken several forms in the past few decades. One trend, occurring predominantly in the late 1960s and early 1970s, involved considerations of mandatory sterilization. Another, and more recent, is the effort to legally coerce specific prenatal behaviors.

Deliberations over mandatory sterilization had their beginnings during the eugenics era. The most significant American decision on this issue was rendered by Oliver Wendell Holmes in 1927 in the case of *Buck v. Bell*. The defendant, Carrie Buck, was a young woman institutionalized for mental retardation. She had just given birth to a daughter, and her own mother lived in the same institution; the institution sought to sterilize her. In Holmes's decision, he wrote these now famous words:

> We have seen more than once that the public welfare may call upon the best citizens for their lives. It would be strange if it could not call upon those who already sap the strength of the State for these lesser sacrifices . . . in order to prevent our being swamped with incompetence. It is better for all the world if instead of waiting to execute degenerate offspring for crime, or to let them starve

> for their imbecility, society can prevent those who are manifestly unfit from continuing their kind. The principle that sustains compulsory vaccinations is broad enough to cover cutting the Fallopian tubes. . . . Three generations of imbeciles are enough.[29]

In the ensuing years, Holmes's decision was intermittently implemented.[30] After World War II, the implications of this law were often repudiated owing to its similarity to measures taken by the Nazis. Nevertheless, the ruling remains to this day within aspects of the law, as a controversial option; it has never been explicitly overruled.[31]

Contemporary considerations of reviving sterilization have entered the political arena in the context of genetic duty. Just as Oliver Wendell Holmes sought to protect the welfare of society, so too have modern interventions sought to save state funds, as in the sterilization of welfare mothers and those with genetic disabilities.[32] Such deliberations raised serious concerns in the 1970s. Bioethicist Marc Lappe, for example, argued: "With rare exception, there is, in my opinion, no compelling case for restrictions on childbearing. I am profoundly disturbed by the advocacy of societal intervention in childbearing decisions for genetic reasons . . . or sterilization of those identified as likely to pass on the genetic basis for a constitutional disability."[33] At the same time, women's groups formed active organizations to prevent involuntary sterilizations, and eventually some of the most blatant forms of abuse in the United States were regulated.[34]

More recently, deliberations over sterilization have been replaced with a new version of genetic duty: the effort to *legally* force pregnant women to conform to specific prenatal behaviors. New responsibilities are being enunciated; penalties for violations of these duties have been enforced.

The issue of legal coercion first came to national attention in the 1986 case of Pamela Rae Stewart Monson. According to George Annas, Monson was the subject of "what may be the first criminal charge against a woman for acts and omissions during pregnancy."[35] Monson was charged with failing to obey doctor's orders, which included a prohibition against amphetamines and sexual intercourse, and a recommendation to remain off her feet.

After failing to obey, Monson gave birth to an infant with severe brain damage; the child died after six weeks. Monson was then charged with criminal neglect.

The social issue is, of course, considerably greater than Monson's specific behavior. While there may be general agreement that Monson's actions were remiss, the larger issue is the willingness of the state to monitor and control the behavior of women. The precedent created by Monson's case is now broadening to include an array of behaviors that may be legally enforceable.

Recently, increasingly precise behaviors for a variety of specific ailments have been legally pursued. In assessing the reproductive behavior of women with the illness PKU, for example, John A. Robertson and Joseph D. Schulman write of measures now being considered against "non-compliant mothers." They note that "some women, even though they are informed of prenatal risks and given access to needed services, may still refuse or be unable to comply with the measures needed to avert harm to their offspring. Given the harm that the behavior will cause offspring and the reasonableness of expecting them to act differently, some persons have proposed that coercive measures, including postbirth sanctions and even prebirth seizures be employed when education and counseling fail."[36] In a similar case, writers Thomas B. MacKenzie and Theodore C. Nagel consider appropriate penalties for a pregnant mother with diabetes. These sanctions include enforced hospitalization, "close surveillance," and "3–4 subcutaneous insulin injections per day."[37]

From this emerging perspective, the civil rights of women are viewed as subordinate to their duties to the fetus. Thus a wide range of actions necessary to enforce behaviors are under consideration, including preventive detention, involuntary treatments and procedures, and imprisonment. As Barbara Katz Rothman notes, "We are in danger of creating of pregnant women a second class of citizen, without basic legal rights of bodily integrity and self-determination."[38]

Such a reinterpretation of rights and duties in genetic terms is a potentially dangerous trend in our vision of our general welfare and has critical policy implications. If rights and duties are increasingly defined in terms of genetic fitness, the civil liberties of women, families, and the "less-than-fit" ultimately may be jeopardized. Equally significant, if the genetically "unfit" are then born, despite extensive social and legal precautions, the question of social responsibility emerges. Who will care for those deemed a "burden" to society?

GENETIC BURDENS

A final element in the development of Genetic Welfare, as conceptualized here, is an emerging reassessment of our social responsibilities. Increasingly, we are perceiving the genetically ill and disabled as individuals who should have been prevented—never born. Thus our societal obligations to them appear extraneous, unnecessary, "a burden" we need not bear.

Clinics and special programs for the ill and disabled are much in evidence in our society today. They are a significant part of what we call the social welfare institution, closely embodying its values and methods. Their methods are training and rehabilitation, the enhancement of skills, advocacy of civil rights, and the creation of environments where individuals may function in useful ways.

Today, however, the message is increasingly communicated that physicians, scientists, and to some degree the public view the clients of these services as "accidents," people who should not have happened—because society could have prevented them.[39] The mandate to provide services may be consequently weakening.

The imperative to house, train, and rehabilitate the ill or the different or the "impaired" appears to be declining in favor of a new imperative—genetic prevention and genetic intervention. The burden of the genetically ill is thus viewed as increasingly redundant in a society whose capabilities permit other paths. This has led to a reconsideration of our social responsibilities.

Several years ago, Theodosius Dobzhansky advised this reevaluation of our social burden. He wrote: "Human life is sacred, yet the social costs of some genetic variants are so great, their social contribution so small that avoidance of their birth is ethically the most acceptable as well as the wisest solution."[40]

As Dobzhansky's remarks illustrate, two issues are at stake: first, an assessment of the costs that must be absorbed by the society due to genetic illness and second, an analysis of the contributions of the genetically ill relative to their costs. From this perspective, deliberations of the genetic burden are often posed in terms of "wise investments," "relative needs," and "other priorities."

Financial considerations are of considerable weight in calculating the genetic burden. Garrett Hardin, for example, writes: "The purpose of prenatal diagnosis is to identify individuals before they are born, while the cost of selection is *comparatively* little . . . the aim . . . is to identify a defect at the earliest possible stage so as to eliminate it before the individual and the society have made any appreciable investment in the unwanted genotype."[41] The costs of such births are typically viewed as beyond those that society should properly undertake.

A second element in this orientation is the perceived lack of return on a large investment; thus the contributions of one group, relative to another are often compared. From the vantage point of Genetic Welfare, it seems, we are not all equal, or, in George Orwell's famous terms, some of us are more equal than others.

From this comparative perspective, human worth is weighed and rated; a "bottom line" on human value emerges. As Daniel Callahan observed several years ago: "the introduction of modern cost-accounting and cost-benefit analysis into the genetic equation adds a distinctly different element. We can now, quite literally, put a price on everyone's head."[42] Such calculations, Callahan argued, are unjust, for we typically tend to emphasize the costs of disability, yet fail to perceive potential benefits. We might better ask how society could be reorganized to enhance the relative contributions of the disabled. Further, he argued, if we insist on toting up such figures, we might also ask, "What is society spending on cosmetics this year?"[43]

This devaluation of the ill and disabled over time has evoked a variety of tactical defenses. Leon Kass, for example, reverses the cost accounting. He asks, "how many architects of the Vietnam

War have suffered from Down's Syndrome? Who uses up more of our irreplaceable natural resources and who produces more pollution—the inmate of an institution for the retarded or the graduates of Harvard College?"[44] How one calculates relative worth is, of course, not really the issue. The larger concern is the thought process and the worldview that ensues: the willingness to weigh, rate, and evaluate human beings and then act on this basis.

In the 1980s, perhaps the most extreme version of this trend emerged, formulated by Helga Kuhse and Peter Singer. In the preface to their 1985 book, *Should the Baby Live?* Kuhse and Singer write: "This book contains conclusions which some readers will find disturbing. We think that some infants with severe disabilities should be killed. This recommendation may cause particular offence to readers who were themselves born with disabilities."[45]

Kuhse and Singer follow this dramatic opening statement with the caveat that "nothing in the view we express in this book in any way implies a lack of concern for disabled people in our community." They further note their support for social services for the disabled. Yet such protestations are questionable given their stark recommendations about the lives of the disabled. Kuhse and Singer recommend a period of approximately twenty-eight days after birth, during which a child may be legally "killed."[46] If such actions are, in their view, the most appropriate response, then a defense of disability programs seems at best fainthearted. In no way could this book be considered a defense of the rights of disabled people.

Clearly, if we, as a society, deem the problem of genetic disease as preventable, the contributions of genetically ill individuals as minimal, and the costs of the genetically disabled as exorbitant, we will be unwilling, ultimately, to accept the burden of care. We are now increasingly torn between social and genetic demands. On the one hand, human services do exist; civil rights are being extended. Alternatively, each year, we more vigorously seek to eliminate the very individuals whose lives are being improved. With better screening, better detection, better identification, and better mapping, our ability to stave off the genetic burden becomes more and more accessible. The sheer efficacy of genetic techniques enhances the likelihood of a genetic choice. The inclination to avoid a burden we *can* avoid is now a critical factor in forging our vision of the world.

The three considerations discussed above, the drift toward a variety of modes of *genetic determinism,* the increasing delineation of our *genetic rights and duties,* and the widening effort to reduce the *genetic burden* constitute the precursors of a distinctive worldview. While such assumptions may not yet be pervasive, the core of these ideas is now being promulgated by each new technological development, its media coverage, and expanding use. Their increasing acceptance at a societal level is likely; their ability to alter public policy is probable.

CONCLUSIONS

Throughout our history, scientific and technological developments have profoundly altered our social organization and consciousness. From the invention of spectacles to the dissemination of the printing press, scientific and technological developments have periodically transformed our adaptation to the physical world. Our newfound abilities in genetic and reproductive engineering are thus but one more step in a long line of events in technological history. But like all such phenomena, these developments have latent consequences; they evoke changes in our worldview and our social behavior.

According to Thomas Kuhn, scientific revolutions cause us to "see the world . . . differently" so that "we may want to say that after a revolution, scientists are responding to a different world."[47] Developments in genetic and reproductive engineering are of this significance, equivalent to the far-reaching revolutions of the past. In 1977 Craig Ellison wrote of these new biological developments: "A revolution is underway. Its potential effects have significant implications for individuals, families, and whole societies. Its ultimate impact could be more pervasive and dramatic than that of the Copernican, Industrial, and Darwinian revolutions. Like these, it is the product of scientific investigation and technological application."[48] In 1987 Linda Bullard echoed these sentiments. "If the present trend continues, genetic engineering will very soon permeate every facet of human activity, having profound social, political, legal, and economic ramifications. Reproduction technologies represent one of the more visible manifestations of the new industry. Perhaps more subtle, but not less dramatic, will be applications in agriculture, pharmaceuticals, animal husbandry, energy production, pollution control, and the military, to cite only a few."[49] Few doubt the momentous changes that are to come.

As we enlarge the boundaries of our technological abilities in genetic and reproductive intervention, we begin to create new choices for ourselves. Such decision making, once beyond our capacities, is now within our hands. To some extent, we can determine the acceptability of human lives, existing and potential. The critical issue, then, is how wisely we will make such decisions. On what basis will decision making occur?

If we view the source of human behavior in increasingly biological terms, we will inevitably shift the solution of social problems to individuals, thus diminishing the significance of social intervention. If we increasingly assess rights and duties in genetic terms, we may ultimately abrogate civil liberties and social rights, thereby jeopardizing fundamental freedoms. Finally, if we choose to calculate the price of certain lives, then deem them too costly, we may be unwilling to undertake the responsibilities of care, thus ceding what many now consider an essential societal responsibility: the support and care of the ill and disabled. If all these trends continue, the vision of Genetic Welfare will become a significant contemporary perception of the world; the worldview of Social Welfare will be consequently diminished.

Social changes of this order do not occur quickly. They touch our lives, slowly, incrementally, nudging out older views and imperceptibly replacing them with new ways of looking at the world. In the following chapter we will consider the process by which genetic considerations may seep into our worldview.

2

A Hidden Agenda in a Technological Age

pity this busy monster,manunkind,

not. Progress is a comfortable disease:
your victim(death and life safely beyond)

plays with the bigness of his littleness
—electrons deify one razorblade
into a mountainrange; lenses extend

unwish through curving wherewhen till
unwish
returns on its unself.
A world of made
is not a world of born—pity poor flesh

and trees,poor stars and stones,but never this
fine specimen of hypermagical

ultraomnipotence. We doctors know

a hopeless case if—listen:there's a hell
of a good universe next door; let's go

—E. E. CUMMINGS,
"pity this busy monster,manunkind."

Since Hiroshima-Nagasaki, we have entered into an increasingly uneasy alliance with science, technology, and progress. The metaphorical images are recurrent: Eve, Faust, Prometheus, Pandora, Jekyll-Hyde, Dr. Frankenstein. The emptied face of J. Robert Oppenheimer could be added to the list. "Physicists have known sin," he confessed to the world.[1]

The allusions typically suggest bargains or exchanges in which individuals or their descendants lose something—their innocence, their integrity, sometimes their lives. A pattern of vulnerability is professed. The perception is frequent that science may take us in directions we have not chosen, impose decisions on us for which we are not prepared. Disquieting to many of us is the way science and technology often infiltrate our lives. Typically, we do not know new technologies are developing. When we learn of them, it is too late to turn back. The techniques are in our hands; we must be prepared to deal with their consequences.

The biological revolution has entered our societal consciousness in this cursory fashion. The multiple steps toward accomplishment of a technique typically occur in the quiet confines of the laboratory. Technologies then make their appearance fully realized or far down the path of achievement. Although often accompanied by an initial phase of interest and excitement, most are quickly absorbed into our technological arsenal. They seem to melt into our lives, but another facet of the capabilities and dilemmas we take for granted in contemporary society.

This situation is reflective of a broad trend in public life occurring over several decades: a shift from intense public disputes over disparate positions to a quiet process of technical introduction.[2] In this way, many public policy outcomes are products not of public debate but of subtle administrative maneuver. This process is the subject of this chapter.

BY FIAT, NOT PASSION

In a now famous observation on public policy, President John F. Kennedy once noted: "the fact of the matter is that most of the problems . . . we now face are technical problems, are administrative problems. They are very sophisticated judgements that do not lend themselves to the great sort of passionate movements which have shaped this country in the past."[3] Developments in genetic and reproductive engineering fit Kennedy's description. As Max Weber might have put it, biology has become bureaucratized—sine ira et studio—without hatred or passion.[4]

The emergence of Genetic Welfare, unlike the "passionate movements" of the past, is a quiet revolution insinuating itself into everyday life in incremental fashion. Decision making is by fiat; accomplishment of a technique renders it a part of our world. The introduction of new techniques is but another item on the hidden agenda of our technological age.

With only rare exception, genetic and reproductive technologies have entered the paths of routine technological development, part of the inexorable forward march of technique, described in classic form by Jacques Ellul: "technique . . . progresses by means of minute improvements which are the result of common human efforts and are indefinitely additive until they form a mass of new conditions . . . what is decisive is this anonymous accretion of conditions for the leap ahead."[5]

Such technique, Ellul notes, is "self-augmenting"; it multiplies upon itself as a basic condition of modern life, accruing in small cumulative steps. Although such change is rapid by the criteria of past centuries, it is relatively innocuous in the context of all other social change now surrounding us.

Technique often occurs under a film of invisibility that is confounding to the modern observer. As Leon Kass remarks, "Introduction of new technologies often appears to be the result of no decision whatsoever, or of the culmination of decisions too small or unconscious to be recognized as such."[6] Through such "nondecision making," we shift inevitably forward toward the next nondecision.

In a classic axiom on the budgetary process, Aaron Wildavsky once observed that "the largest determining factor of the size and content of this year's budget is last year's budget."[7] Much the same could be said of genetic and reproductive technology—that last year's advances determine this year's research. Thus, in cumulative and additive steps, through administrative and technical decisions, the biological revolution has come to be part of the contemporary landscape.

In the arena of biological technique, we seem to be experiencing what economist Charles Lindblom once called "endless nibbling." Lindblom critiqued modern public policy for failing to take a "good bite," yet admitted "this kind of persistence in policy-making has transformed the society." The United States, Lindblom wrote, "has gone through an industrial revolution, an organizational revolution, a revolution in economic organization . . . a revolution in the role of the family—but all through policy sequences so undramatic as to obscure the magnitude of the change."[8] Like Lindblom's process of "endless nibbling," the biological revolution is eroding its way into public policy and social consciousness.

Accomplishments are now routinely communicated to the public through the mass media. Each development becomes another item in the daily news summary. Often, the refinements of these techniques, their progress, and their multiple applications quickly follow. Thus there is little opportunity to discuss first questions when techniques are so far along.

Further, there is no single locus for the development of genetic and reproductive research, making its source difficult to identify and its control harder to envision. Internationally, the biological revolution is occurring in thousands of different universities, hospitals, corporations, and laboratories. Unlike research on the atomic bomb which, for better or worse, was in the hands of a small number of people, biological investigations are now pursued in a vast number of settings and through the work of thousands of physicians and scientists. This dispersion of genetic and reproductive research makes no single unit the responsible agent; each agency sets its own pace and its own objectives.

The uses and applications of technologies are similarly dispersing in public use and public consciousness. One obvious example is genetic screening. As originally envisioned, genetic screening was designed for the exceptional case: women thirty-five years of age and older, and those with problematic clinical histories. While amniocentesis may still be used sparingly, newer and simpler tests, such as alpha fetoprotein (AFP) analysis, are more widely used on younger women and those without anticipated genetic problems. Further, AFP screening "is the first prenatal screening technique mandated by law." California now requires obstetricians to inform their patients of the availability of the technology. The method also has been "strongly recommended by the American College of Obstetricians and Gynecologists."[9]

The President's Commission for the Study of Ethical Problems has predicted a major increase in genetic screening in the future: "Before the end of this century . . . genetic screening and counseling are certain to become major components in both public health and individual medical care . . . the time can already be envisioned when virtually all relevant information about a person's genotype including all his or her "abnormal" genes and chromosomes will be readily accessible."[10] The relatively recent technique of genome mapping is now similarly seen as a candidate for widespread use. George Annas notes: "One can imagine a time when your entire genome could be mapped at birth. . . . Your genetic profile could be used for the rest of your life to determine who you marry, what jobs to apply for, and for insurance companies to decide if you are a good risk."[11]

A similar example of increasing use and broadening social acceptance is in the development of in vitro fertilization. Louise Joy Brown, the world's first "test-tube baby," was born on July 25, 1976. At the announcement of her birth, extensive media attention was devoted to the implications of this new technology.[12] Yet the publicity soon faded. Today, thousands of IVF procedures are being performed throughout the United States and the world; the number of IVF clinics is rising yearly.

Small towns and community hospitals across the country now proudly announce the establishment of IVF clinics.[13] In vitro fertilization is referred to as one more "treatment option"[14] in the range of available technologies; its integration into our world is becoming complete. Problems raised are now directed only to special cases. Technologies once seen as extraordinary are coming to be accepted for general use.

Even exotic technologies such as embryo freezing are being routinized in this fashion. In 1988 Andrea Bonnicksen noted,

"Only four years ago, embryo freezing was considered a technique raising 'disturbing,' 'extremely difficult,' 'incredibly complex,' and even 'nightmarish' ethical issues. Currently, however, at least 41 of the 169 infertility clinics in the United States have added freezing to in vitro fertilization."[15] Rapidly, the ethical issues have been viewed as resolved, fundamentally because the practice is considered commonplace.

When technologies are routinized in this manner, the issues that arise become practical questions of implementation. Physicians and scientists ask: "How can we better deliver this service?" "Can we increase our clientele?" "How can we improve this or that phase of the procedure?" Under such conditions, we, as a society, are less likely to ask "Do these technologies represent directions we wish to take?" and more likely to ask "How best shall we tinker with this or that part of ourselves?" In so doing, the foundation of the argument shifts, for fundamental issues of values and direction are avoided in favor of implementation. Policy alternatives are reformulated, not by decision but by being edged in the direction of existing technique. The tenets of Genetic Welfare are at hand, not because we have necessarily determined they constitute the best means to a desired end, but because the instrumentality to accomplish them is so readily accessible.

Increasingly, issues that suggest the need for serious debate at their onset are regularized by the sheer familiarity of their practice. If we *do* it, we *understand* it; if we understand it, we *accept* it; if we accept it, we *use* it. Its widespread dissemination is ensured. Finally, it is absorbed into the fundamental institutions of our society.

The nature of this contemporary process is thus distinctively different from the intense and "passionate" debates of the past. Biological movements have occurred throughout history, but they have never been so accessible technologically, nor so accretive administratively. As a contrast in the *process* of public policy (though not necessarily the substance), it is useful to consider briefly the eugenics movement of the turn of the century. Its fundamentally different means of introduction suggests the nature of the policy change that is upon us.

EUGENICS: "A PASSIONATE MOVEMENT"

The eugenics movement of the turn of the century flourished first in England, then in the United States, and finally in Nazi Germany. Eugenicists sought to increase the birthrate of the "fit" and decrease the productivity of the "unfit." To promote these ends, the English founder of the movement, Francis Galton, advocated programs for the "select," including housing subsidies and patronage by the wealthy.[16] To Galton, the mission of charity was to "help the strong rather than the weak; and the man of to-morrow rather than the man of today."[17] Eventually Galton promoted eugenics as "an enthusiasm . . . so noble in its aim, it might well give rise to the sense of religious obligation."[18]

Galton's visionary aims attracted a wide following, including George Bernard Shaw, one of many converts to eugenics, who argued that "there is no reasonable excuse for refusing to face the fact that nothing but a eugenics religion can save our civilization from the fate that has overtaken all previous civilizations."[19] The passions of this movement deeply stirred the social and intellectual communities of Europe. By the turn of the century according to Marc Haller, eugenics had become "a world movement," with centers in Germany, Scandinavia, France, Austria, Italy, Japan, South America, and the United States.[20]

The eugenics movement was received with particular fervor in the United States. In the early 1900s the threads of progressivism, naturalism, and xenophobia combined to make eugenics a profoundly enticing mode of thought and action.[21] According to Kenneth Ludmerer, "To most eugenicists, the movement was not just a social crusade, but a moral crusade as well. . . . In this tradition, American eugenicists spoke of 'the religion of evolution,' 'the duty of upbuilding the human race,' and 'the moral implication in the doctrine of evolution.' . . . The devotion of most eugenicists to the moral crusade of eugenics was limitless. . . . When they campaigned for legislation, officials and other citizens could not help but heed the fervent, impassioned pleas of many eminent persons."[22]

American eugenicists such as Margaret Sanger exhorted the nation to "breed more children from the fit, less from the unfit" as a means of fending off the destruction of civilization.[23] Theodore Roosevelt was another figure who responded to the call of eugenics. His determination to halt "race suicide"[24] initiated an emotional and intense movement to regulate reproductive patterns in the United States.

The millions of immigrants then entering the United States were the primary objects of attack. The eugenics movement called for mandatory sterilization of those deemed unsuitable to bear children and successfully secured the Federal Immigration Restriction Act of 1924.[25] The nation reached one of its great peaks of ethnic and racial hostility during the congressional hearings on this act. Between 1921, with the passage of a temporary act, and 1924, with the creation of a more permanent policy, according to Marc Haller, "American racism reached its climax. Testimony before Congress, political speeches, and numerous books and articles carried the concepts of Nordic superiority and race purity to the public. . . . When Congress formed a permanent policy in 1924, racial considerations were foremost, and every Congressman became his own eugenist."[26] This xenophobia held sway in the United States until Europe once again began to adopt eugenics, this time in combination with an emerging fascism.

In the 1930s Nazi Germany took over the eugenics frenzy. By linking eugenics with historic visions of Teutonic supremacy, the Holocaust emerged. This "passionate movement" for biological supremacy included "euthanasia" of those judged mentally or physically unworthy, Aryan mating camps, and ultimately, of course, mass genocide.

In sharp contrast to the eugenics movement of the early twentieth century, contemporary biologism is neither religious nor messianic in tone. Today, unlike the original advocates of eugenics, one need not be a "true believer" to accept the tenets of Genetic Welfare or even newly defined eugenic goals.[27] It is not fervor that is emerging as a problem to critics, but indifference. The advances of the biological revolution are perhaps too ordinary and too routine a part of our everyday technical and administrative decisions.

One underlying reason for this ready routinization of genetic and reproductive engineering is that most of the technologies of today's biological revolution are advanced through the institution of medicine; as such, they have the sanction of health care. Techniques are typically posed not as efforts to change the world or civilization or ourselves, as was the case with eugenics, but as a means to improve our health and well-being. The legitimacy of these techniques is thus far more assured; potential problems are more difficult to pose.

THE MEDICAL DIMENSION AND BEYOND

Today, virtually every accomplishment of genetic and reproductive engineering, applicable to human beings, has been mediated through the institution of medicine; each progression has been viewed as palliative or ameliorative, preventive or curative, as beneficent as penicillin or the polio vaccine. The biological revolution, in great part, has made its appearance in the form of medical care. Like many other social phenomena of late, genetic intervention has been "medicalized."[28]

At first glance, the placement of genetic modifications in the medical arena seems fitting and appropriate. Clearly, most of the new technologies deal with fundamental medical issues such as reproductive disabilities and hereditary genetic disorders. Yet this placement may also be debatable, for new genetic technologies also have the capacity to transcend medicine and enter new arenas of intervention and definition.

To Robert Francoeur, for example, "genetic engineering and our reproductive technology have catapulted the physician far beyond remedial medicine into the domain of creative designing and positive genetic planning."[29] Similarly, Amatai Etzioni asks, "Do we not, by opening the door to one kind of genetic engineering—intervention into disease—also open another door to improvement of the race, the door we want closed?"[30]

If genetic and reproductive technologies are routinely transmitted to the public through the institution of medicine, they are likely to be widely accepted. Medicine, with its mantle of legitimacy, has already contributed substantially to public acknowledgment of these technologies. If, at the same time, these advances go beyond medicine into other arenas, these arenas, too, may come to be widely accepted. New assumptions may be introduced, new behavioral patterns become commonplace. Genetic interventions may become as standard as the yearly medical checkup. In this way, the institution of medicine may serve as a conduit for routinizing a new social philosophy, one that may fall well beyond medical spheres.

According to many social historians, medicine served just such a conduit function in the late nineteenth century, by defining the weaknesses of women and determining their incapacity to function in the public world. Then, as now, medicine served as a mechanism for the transmission of fundamental social attitudes. Barbara Ehrenreich and Deirdre English explain: "Medicine stands between biology and social policy, between the mysterious world of the laboratory and everyday life. It makes public interpretations of biological theory. . . . Biology discovers hormones; doctors make public judgements on whether 'hormonal imbalances' make women unfit for public office. More generally, biology traces the origins of disease, doctors pass judgement on who is sick and who is well."[31]

In the nineteenth century, physicians declared women incapable of public work. The judgment of "female invalidism" led physicians to deter middle- and upper-class women, in particular, from burdensome efforts of all kinds.[32] A regular regimen was prescribed of rest, protection from the elements, insulation from stress, and a coddling of the spirit. According to Richard W. Wertz and Dorothy C. Wertz, "Doctors devised a vast litany of specific dangers to women's health. . . . The effect of their increasingly urgent and widespread advice was to teach women that their biology was . . . so sensitive that they must live a cautious life shaped by the demands of their inward parts. Women were warned . . . that they must keep to the social roles of wife and mother for which their biology was fitted and in which they were safe. To venture beyond the domain of the home was to risk danger. . . . The world of intellect and business was too full of nervous shocks for women to enter it without endangering their physical being and personal character."[33]

Wertz and Wertz aptly note that "doctors did not invent the nineteenth-century cultural views about women's proper place and behavior," but did serve to convey them.[34] Their research and science supported these views, and they were in a position to dispense them with authority. The institution of medicine is often in this position of authority. It is a critical leverage point where new societal assumptions may be introduced. In our era, as in the nineteenth century, medicine may also serve as a means to define and redefine fundamental social and political issues.

The arena in which medicine has most successfully transmitted new values of "genetic fitness" is through its advocacy of genetic screening. Today, virtually all obstetricians must consider its advisability, and such genetic considerations are moving substantially beyond the speciality of obstetrics. Even primary care physicians are being "urged" to convey genetic information. Dr. Mack Lipkin, Jr., director of primary care at New York University, argues, "Primary care physicians are the key to effectively introducing most people to genetic counseling and screening."[35] Thus the profession of medicine is quickly seeking to make genetic issues broadly relevant, at the entry level–primary care posts of medicine. Questions of genetics are moving out of the specialty laboratories and into the broad intersecting points of medicine, much as they did in the nineteenth century.

Another arena in which the dividing line between medicine and new social perceptions is graying is in the development of recombinant DNA, or what is often called gene splicing. Many current applications of this technique are clearly medical and involve the creation of treatment substances such as interferon, insulin, somatostatin, thymosin, and urokinase.[36] At the same time, recombinant DNA also permits the creation of "species hybrids," which may serve multiple human needs and purposes.

In the past decade, several new organisms have been developed through recombinant DNA techniques, such as an oil-absorbing bacterium used in cleaning industrial oil spills and an organism designed for agricultural use which alters the freezing temperature of crops. A patent was granted in 1980 for the oil-absorbing bacterium, representing the first patented "life form" in history.[37] In 1988 the first higher life form patent was granted for a mouse, "genetically altered" to make it susceptible to cancer-causing chemicals; many more such patents are planned for the future.[38]

Thus the technology of recombinant DNA along one path generates medical treatment substances, along another path, new organisms. Yet the two outcomes merge at the conceptual level. Each is for human ends. Both medical treatment and potential life form creation become fundamentally the same; the patents suggest our willingness to equate them. Moreover, the patenting process has once again come about in a routine and undramatic form. Of the first patent case, *Diamond v. Chakrabarty*, Harold Green notes, "Chakrabarty did not involve any burning issues of public policy. The only issue presented in the case was a mundane legalistic question. Did a statute presently in force authorize the grant of a patent to a new bacterium invented by Chakrabarty?"[39] In this "mundane" way, new social assumptions may be tied to new medical approaches.

As a final suggestion of slippage from medical to social arenas, we may consider the new medical treatment of tissue transplantation. Researchers have recently made strides in the relief of disease through the procedure of fetal tissue implants. Parkinson's disease, for example, may be modified by this process. Tissues for transplant are currently "harvested" from miscarried or aborted fetuses. Soon, however, this "harvesting" may go further. The question has arisen, shall we *create* a life, through reproductive technologies, specifically for the purpose of implantation treatment? For example, we could artificially inseminate a woman in order to abort the resulting fetus which would then be used for treatment purposes. This practice has not yet begun, but it is being contemplated.[40] Here again, a search for a medical treatment, aided by the technology of artificial insemination, yields as well a reassessment of human life, in the name of medicine.

The legitimacy medicine offers to genetic and reproductive technologies is an important element in the emergence of Genetic Welfare. The institution of medicine has served to introduce many new technologies to the public. With these technologies come a set of assumptions, about social life and the human community. Such a process is typically slow, quiet, and cumulative. These orientations then become part of the most common aspects of our lives; a worldview is generated in the process.

BUSINESS AS USUAL

Beyond the medical dimension of genetic and reproductive technologies lies their relationship to business and profits. As with most drugs, medical treatments, or medical hardware, genetic and reproductive engineering technologies are being marketed for considerable or anticipated profits.[41] This, too, contributes to the accretive nature of Genetic Welfare.

If a product is sold on the market place, it becomes part of the available resources of a society. It may become standardized, a familiar and commonplace commodity. Moreover, it becomes institutionalized into the economic structure. Companies then have a vested interest in maintaining its viability, continuance, and use.

The extraordinary investments in genetic engineering firms are now well known. When two of the earliest firms, Genentech and Cetus, first went public, they garnered some of the largest investments in the history of the stock market.[42] Other companies have done almost as well. Moreover, the rapidity with which basic research has entered the commercial sphere has startled many observers. In 1981, for example, then congressman Albert Gore noted that in genetic engineering "there seems to be no phase of applied research: the discovery of the basic scientists may go directly and swiftly from the laboratory bench in the university into a profit-making venture."[43]

Today, in the United States, there are several hundred genetic engineering firms, with new ones accumulating every year.[44] Internationally, there are several hundred more. Reproductive technologies are similarly linked to profit-making firms. Fertility and in vitro fertilization clinics are multiplying, their treatments being sold at high cost to their clientele. Surrogacy, whatever its ultimate legal status, has been a profit-making venture.

Universities are intimately linked to these profit-making ventures, as well, and further serve to legitimate the research. By 1984, "nearly half of all biotechnology companies in the United States provided some kind of funding for university based genetic engineering research."[45] Other universities, such as the Massachusetts Institute for Technology, have set up independent profit-making research centers. The linkages between research and distribution are now increasingly interwoven and serve to intensify the diffusion of new technologies.

The entry of genetics into the world of high finance parallels the "corporatization of medicine" phenomenon of the last decade. Paul Starr notes: "The rise of a corporate ethos in medical care is already one of the most significant consequences of the changing structure of medical care. It permeates voluntary hospitals, government agencies, and academic thought as well as profit-making medical care organizations. Those who talked about 'health care planning' in the 1970s now talk about 'health care marketing.' Everywhere we see the growth of a marketing mentality in health care."[46]

Starr argues that this shift will have serious consequences for the distribution of health care which are "likely to aggravate inequality in access," thus jeopardizing the health care of the poor. He notes: "Profit-making enterprises are not interested in those who cannot pay. . . . A system in which corporate enterprises play a large part is likely to be more segmented and more stratified."[47] Such outcomes are likely to be reflected in the corporatization of genetics as well. The inclusion of genetics, like medicine, in the corporate marketplace may serve to promote the most profitable ventures, rather than the most therapeutic, with less attention to equity or long-term implications. As Walter Gilbert, "a former Harvard professor and founder of Biogen, one of the early bio-technology firms," notes, his company will only pursue what will be of value "next year or three years from now, not something of social value twenty years from now."[48]

The integration of genetic and reproductive technologies with the business community gives genetic technologies a foothold in the most significant structural edifice of American society, its economy. Goods and services sold at profits become difficult to dislodge and challenge; cigarettes are but one example. The expansion and elaboration of the "medical-industrial complex" into new spheres of genetic and reproductive technologies may thus serve as further encouragement for their acceptance and justification for their use. Their corporatization makes a worldview accompanying them a more likely eventuality.

THE INSTITUTIONALIZATION OF A GENETIC APPROACH

The gradual absorption of new technologies into our institutional structure, their acceptance with little public debate, the medical aegis that legitimizes new procedures, and the commercialization of technologies all serve to bolster the development of Genetic Welfare as a worldview. As noted, this process has been, on the whole, prosaic and routine, rather than dramatic, a familiar phenomenon of everyday life. The institutionalization of social attitudes typically occurs in this fashion.

The institutionalization of discrimination is an example of this process. Several years ago, in their book *Institutional Racism in America*, Louis Knowles and Kenneth Prewitt drew a distinction between individual racism and what they called institutional racism.[49] To Knowles and Prewitt, institutional racism was embedded in the fabric of society, dwelling in the interstices of our social institutions. Beyond "Jim Crow" laws or formal systems of apartheid are simply the modus operandi of a society and the routine functioning of its social institutions which generate the practices of racism. It is the world as we know it, buttressed by its social practices and circumscribed by its dominant ideology.

Institutional racism reflects a worldview that fosters discriminatory practices, but it is a hidden set of assumptions, difficult to discern. As Kristen Lucker notes, "An interesting characteristic of a worldview . . . is that the values located within it are so deep and so dear to us that we find it hard to imagine that we even have a 'worldview'—to us it is just reality."[50]

Peter Berger and Thomas Luckmann, in their book *The Social Construction of Reality*, offer a similar concept in their discussion of a "symbolic universe." According to Berger and Luckmann, a symbolic universe is a vision of the world, generated by our social institutions, which legitimates our behavior. We come to see our beliefs as "having a life of their own," "the world is as it should be." Such legitimation "justifies the existing institutional arrangement by giving a normative dignity to its practical imperatives."[51] In developing a symbolic universe, we create order and give meaning to our way of life.

Societies, according to Berger and Luckmann, "are constructions in the face of chaos."[52] More important, they are "social" constructions, and science, they argue, may be the ultimate constructive force of the modern age. Past symbolic universes may have derived from mythology, theology, and philosophy; but today they emerge from science and technology. In contemporary society, increasingly, it is science and technology that pose and then answer the "ultimate questions."[53]

Perhaps one acknowledgment of this alteration in symbolic universe is the frequent use of mythological, philosophical, and theological metaphors in discussions of the biological revolution. Just as critics raise ancient metaphors to protest the progress of science, so, too, scientists themselves refer to their own creations as "chimeras," "minotaurs," and "cyborgs"—half-man/half-beast or half-man/half machine.

Physicians and scientists do seem to be aware they have the capacity to define and redefine the ultimate questions. As Robert Francoeur has written, "we face the possibility of so altering our human constitution that a new creature may be born, which despite its human origin and gestation, we may not consider a legitimate member of the human species."[54] Anthropologist Robert Redfield once suggested this distinction—between "human and not-human"—was one of the central components of a worldview.[55] Genetic Welfare may soon require this distinction.

CONCLUSIONS

The introduction of genetic and reproductive technologies has occurred, like many other aspects of our modern world, in a technical and administrative fashion. Unlike the "passionate movements" of the past, the momentum of the biological revolution has been generated by a technological imperative which has given it a cumulative and accretive dimension. New technologies appear rapidly, and there is little effort to stop them. Their legitimacy is granted by medical appellation; they have been further sanctified by their integration with the marketplace. Genetic and reproductive technologies are being quickly absorbed by the fundamental institutions of the society.

Unlike eugenics, Genetic Welfare combines a new technological capacity with a new sociological base. It affects the mass of society through an increasingly accessible technology that promises to become more available with each passing year. New developments are forwarded to the public through the mechanism of mass communications, reinforcing a genetic worldview on a routine and systematic basis. Each new technology claims to offer a new means to assuage our physical and social difficulties.

Genetic and reproductive technologies offer a new route to promote the general welfare. These technologies permit us to define life in new ways. Our expectations of who we are, what constitutes human well-being, and what tragedies we need tolerate shift in the process. Genetic and reproductive technologies offer a new means to "construct order out of chaos"; a new worldview becomes possible.

In the following chapters we will consider the impact of Genetic Welfare on specific subgroups in our society, groups which by virtue of their current peripheral position are vulnerable, powerless, or alienated. This approach will be deliberately problem-seeking. The potential dangers for those who are already outsiders in our society will be assessed.

PART II

Assessing Genetic Welfare

Part I of this book began an exploration of the worldview of Genetic Welfare. We examined the idea that genetic concepts are coming to subvert more socially oriented perspectives. New assumptions are being generated: biological determinism, genetic rights and duties, and genetic burdens. These assumptions are entering our worldview slowly and imperceptively, changing our decision-making processes and altering our behavior. Basic social institutions, such as medicine and business, enhance the absorption of genetic ideas by transmitting them to ever widening circles of the public.

Part II examines the consequences of a genetic worldview for three social problem areas: stigma, powerlessness, and alienation. For each social issue, we will consider the nature of the progress that has been made and how this progress may be lost. The overriding question, in each chapter, is how the emergence of genetic technologies may offset recent social advances. What new forms of stigma will be produced? What new inequalities will be created? What new types of medical alienation will be generated?

The essential concern of Part II is that certain groups, already vulnerable or disadvantaged, become targets for new forms of discrimination or disadvantage. Although genetic technologies may aid some groups whose members have particular physical conditions, they have latent effects on other groups and on society as a whole. Genetic technologies have the power to define new types of diseases, labeling individuals never before labeled or reinforcing the intensity of old labels. Such technologies also constitute a new social resource, available to some but not to others. Likewise, genetic technologies have the potential to create new modes of medical intervention, shifting the nature of medical practice. It is vital that we consider the costs that may be created by genetic technologies as well as their benefits. To anticipate such problems is to perhaps be prepared to mitigate their negative consequences.

3

The Problem of Stigma

As an intelligent woman, but handicapped by blindness, I do not in the least object to the classification which has associated us with criminals and feebleminded. . . . [But] I would remind you that sixty percent of the blind sent out from schools are self-supporting. . . . When I observe the idle, selfish, shallow sons and daughters of the rich, spending their days in worthless pursuits, making no contribution of life and service for society, no answer to the great cry of humanity, I ask myself the question—who in the sight of God are the unfit?

—MISS ADAMS of Cleveland, Ohio, at the National Conference of Charities and Corrections, in response to a paper entitled "Eugenics and Charity," 1912

The most chilling thing said this afternoon is that we couldn't afford, or ought not to permit, another potential Dostoevski to come into existence because one could early discern the epilepsy. . . . Are we really so sure that the qualities of normality with reference to some biological norm are the kinds of qualities that are going to make life interesting, rich and fulfilling?

—JAMES GUSTAFSON, at the "Symposium on the Identity and Dignity of Man," sponsored by Boston University and the American Association for the Advancement of Science, 1973

Discrimination against those deemed "unfit" has been a powerful force throughout history. In recent decades, however, new social policies have attempted to reverse the historical pattern of stigmatizing disability and deviance. To this end, there has been an increasing attentiveness to issues of mainstreaming, deinstitutionalization, civil rights, and the problems of labeling. The potential impact of Genetic Welfare on these new polices constitutes the central focus of this chapter.

Several dimensions of social change will be considered: Will new modes or categories of stigma be introduced by genetic interventions, further segregating those defined as "normal" and "abnormal?" Will policies emphasizing social integration and civil rights be weakened in favor of programs stressing birth prevention? Finally, will Genetic Welfare diminish the societal tolerance of physical, mental, and social differences?

THE SOCIETAL RESPONSE TO DEVIANCE AND DISABILITY

The acceptance of deviance and disability poses a continuous challenge to the fabric of our social structure.[1] The boundaries between the "normal" and the "abnormal," the "fit" and the "unfit" symbolize our standards of social tolerance. When we soften these boundaries, we acknowledge as "normal" or "fit" a greater number of human types and social behaviors. When we harden these boundaries, the outgroups grow larger and more numerous; treatment is often harsher.

Who we view as deviant is intimately connected to our understanding of *why* deviant behavior occurs. Problems of deviance and disability often have been understood in biological terms.

People were simply "born that way"—irremediably—as misfits and criminals. In the early 1900s, for example, Caesare Lombroso founded the first "scientific" field of criminology, suggesting broad biological principles for the evolution of criminal types. Eugenics followed suit: tramps, beggars, and paupers, those who were blind, deaf, and disabled, murderers, and rapists, were grouped under immutable definitions of unalterable destinies.[2]

Public policy mirrored this biological determinism. By the late nineteenth century, asylums and prisons sheltered the public from the deviants' "harmful ways." It was believed that little could be done to alter their functioning; the potential for rehabilitation was viewed as minimal. Warehousing was considered the most acceptable alternative, and harsh punishments were thought appropriate to keep people in line.[3]

We have progressed beyond this biological fatalism only very slowly and very arduously. It is one of the great accomplishments of the twentieth century, therefore, to have brought about significant change in our perception of deviance and disability. One of the most important of these advances is our effort to view these issues in social as well as biological terms.

In this new social approach, deviance and disability are viewed as part of a relationship: an interaction formed as much by the perceptions and actions of society as by the deviant. The process that initiates this transition from "normal" to "abnormal" or "fit" to "unfit" is often called stigmatization or labeling. Efforts to prevent such labeling and its stigmatizing consequences represent important contributions to modern social policy. In the following sections we will assess the directions of contemporary labeling and stigma theories, and consider how the tilt of Genetic Welfare may affect these advances.

STIGMA AND LABELING: AN INTERACTIVE PERSPECTIVE

Since the 1960s, extensive research has been conducted on the societal responses to deviance and disability. One of the earliest of these works was Erving Goffman's landmark study, *Stigma*.[4]

Goffman viewed the process of becoming a deviant as an interaction between those he called the "discreditable" and those who do the "discrediting." Goffman argued that "a language of *relationships,* not attributes is really needed," for the societal stamp of disapproval may ultimately generate more harm than any trait itself.[5]

> By definition . . . we believe the person with a stigma is not quite human. On this assumption we exercise varieties of discrimination, through which we effectively, if often unthinkingly, reduce his life chances. We construct a stigma-theory, an ideology to explain his inferiority and account for the danger he represents, sometimes rationalizing an animosity based on other differences such as those of social class. We use specific stigma terms such as cripple, bastard, moron in our daily discourse as a source of metaphor and imagery. . . . We tend to impute a wide range of imperfections on the basis of the original one.[6]

Howard Becker, in his classic study, *The Outsiders*, makes the same observation in regard to deviance: "From this point of view, deviance is *not* a quality of the act the person commits, but rather a consequence of the application by others of rules and sanctions to an 'offender.' The deviant is one to whom that label has successfully been applied; deviant behavior is behavior that people so label."[7] More recently, Edwin Schur has argued the same point of view: "it is *through* social definitions, responses, and policies that particular behaviors, conditions, and individuals acquire their 'deviantness'."[8]

In the fields of health and mental health, this interactive perspective has come to be highly influential. For example, there has been much recent attention to problems of language and terminology.[9] To define people by the status of their disabilities, it is argued, causes a stigmatizing mind-set in the social audience. Thus there has been a new emphasis on noting that people *have* disabilities, rather than *are* their disabilities (e.g., the blind, the deaf). Similarly, the notion that a "handicap" is not a *personal* attribute, but rather a *social* obstruction (e.g., a lack of curb cuts on sidewalks), addresses the significant interaction between language, social structure, and stigmatization.

Jane Mercer, in her book *Labeling The Mentally Retarded*, clearly illustrates this perspective. "Whom we call mentally retarded, and where we draw the line between the mentally retarded and the normal depend upon our interest and the purpose of our classification. The intellectual problem of mental retardation in the community is, ultimately, a problem of *classification and nomenclature*. . . . Mental retardation is not a characteristic of the individual, nor a meaning inherent in his behavior, but a socially determined status, which he may occupy in some social systems and not in others, depending on their norms."[10] The condition of mental retardation, according to Mercer, is a variable trait; what is significant is its social context and how the label is applied.

In all, these interactive perspectives argue that problems of disability and deviance are "socially constructed."[11] They exist in some settings but not necessarily in others; how we perceive the disability is as critical as the disability itself. Alter the social behavior and we can lessen or perhaps eliminate certain problems. Problems, therefore, become socially remediable. Reduce the stigma, change the environment, create new civil rights, or new modes of perception and we can significantly enhance an individual's life chances.

From these insights, a new view of policy and treatment has emerged. Social programs have been initiated that attempt to prevent or lessen stigmatization, thus avoiding its negative consequences. To achieve this goal, efforts have been made to mainstream and deinstitutionalize individuals who have been considered deviant or disabled. These approaches attempt to make deviance or disability part of the everyday landscape. Through repeated exposure, experience shows, the perception of "differentness" diminishes and fears are reduced; "they" become "us." This has been particularly true in the areas of disability and mental health, and to a lesser extent in the field of criminal justice.

In the fields of disability and mental health, new legislation has facilitated mainstreaming in routine places of work, recreation, and residence. Elevators, ramps, parking spaces, special teachers, and sheltered workshops have been mandated through Section 504 of the landmark National Rehabilitation Act of 1973, the Education for All Handicapped Act of 1975, and the more recent Americans with Disabilities Act of 1990.[12] At the same time, the civil rights of persons with disabilities and illnesses have been more broadly interpreted than they have in the past. New rights of treatment, due process, right of counsel, notice of hearing, "least restrictive alternatives," informed consent, and many other issues have been adjudicated. Many of these interpretations have legitimated and legalized increased access to the larger society.[13]

In the field of criminal justice, juvenile proceedings, in particular, have reflected similar deinstitutionalizing trends. The theory, although not always the practice, has been to protect youth by not making their names publicly known, or by clearing their records, thereby preventing their personal and public identification as criminals. Adult work-release programs similarly represent efforts to mainstream individuals who have committed lesser offenses.

What is common to all these policies is their *social* interpretation—that labels are harmful. Mainstreaming and deinstitutionalization seek to avoid these negative consequences. The affected individuals are acknowledged as potentially contributing members of the society; our shared humanity is affirmed. Inclusion, not exclusion, is sought. These new approaches impute legitimacy to individuals; as remedies, they stress alterations in the environment, not modifications of persons.

These advances, however, despite their great importance, are potentially fragile and vulnerable accomplishments. They have been achieved painstakingly, through lengthy legal intervention and social pressure. Society has yielded only very reluctantly to the acceptance of physical, mental, and social differences. Each step has been a struggle, and setbacks are frequent.

Communities, for example, have often attempted, shamefacedly, but nonetheless vigorously, to prevent "those people" from entering their areas. "Not in My Backyard" has become so frequent a rallying cry, it has been christened a "syndrome" with its own acronym—NIMBY. There are fears that neighborhoods will share in the stigma of an entering group. Accusations of potential violence, disturbances, and even dirtiness are frequent. As Erving Goffman observed, groups frequently embroider a multitude of negative traits on any single problem.

Similarly, monies have been limited for new sites and services. Many of those who have been deinstitutionalized from mental hospitals have been turned out into the streets, homeless and without care, rather than into community residences, as originally planned. Deinstitutionalization became cost-saving rather than lifesaving. Initially good intents were subverted.

In the 1980s, improvements in accessibility were hampered by limited financing. In major cities, bitter controversies over public transportation occurred. Some areas fought to create small fleets of vehicles for people with disabilities, rather than increasing access in their larger systems. Likewise, commercial airlines, as one headline put it, became "Free to Discriminate Against Disabled," based on a court decision limiting access.[14] The provision of special teachers and facilities in public schools also was challenged as too expensive. Specialized centers, classes, and transportation were all cheaper, but the mainstreaming effort suffered; those with disabilities were once again segregated and stigmatized.

How these specific problems ultimately will be resolved remains uncertain. What is clear, however, is that the newness of many of these programs and their high costs make them unstable. They require a fundamental commitment to social change and social justice; without these commitments, such programs are difficult to maintain over the long term. They may be quickly threatened by even subtle changes in worldview.

NEW MODES OF STIGMA AND LABELING

Genetic interventions have the potential to contribute to the formation of new labels and stigmas. With the ability to anticipate genetic disorders and the capacity to know our chromosomal liabilities, we set into motion the very process Frank Tannenbaum once called "tagging, identifying, segregating, describing, emphasizing, making conscious, and unconscious" that is the process of labeling.[15] Bioethicist Marc Lappe has noted this significant potential:

> At the very moment you acquire a "bit" of genetic information about a fetus (or any other person for that matter), you have begun to define him in entirely novel terms. You tell him (and sometimes others) something about where he came from and who is responsible for what he is now. You project who he may or may not become in the future. You set certain limits on his potential. You say something about what his children will be like and whether or not he will be encouraged or discouraged to think of himself as a parent. In this way the information you obtain changes both the individual who possesses it and in turn, the future of that information itself.[16]

Regrettably, what is happening, is that at the precise time social policies are seeking to eliminate, reduce, or mitigate the effects of old labels, new stereotypical images are emerging, based on new genetic information. This enhanced capacity to generate stigmatic situations is likely to harm the overall effort to broaden societal perceptions.

Many of the technologies of genetic and reproductive engineering contribute to labeling, but the most acute issues have developed over genetic screening. James Sorenson, for example, has observed, that "a basic issue which arises in genetic screening revolves about the personal meaning of being labeled genetically aberrant."[17] Similarly, D. S. Halacy has argued that "there are many scientists who argue that screening may lead to the creation of genetic lepers."[18] The image of potential stigma is widespread. Anthony Mazzocchi, for example, has serious reservations about genetic screening in the workplace. He notes, "If an individual is removed from a job because they have some genetic trait, it is like putting a genetic scarlet letter on their forehead. You're dooming them and all their offspring from meaningful work."[19] Concerns over the labeling consequences of screening have been numerous.

One of the new labels recently created through genetic screening is that of "carrier status": individuals with recessive disease-carrying genes. Carriers are not ill, but if paired with another carrier, they may bear offspring with the genetic disease they carry. Sorenson has noted that such labels, in the future, may significantly harm marital prospects and seriously damage self-concepts.[20] Clearly, it can be argued that if genetic "cleanliness" becomes a major preoccupation of populations, the label of carrier status will become increasingly discrediting.

In the 1970s certain social and genetic groups came to be at particular risk for genetic stigmatization. Two in particular were males with an extra Y chromosome (XYY) and African Americans. Both groups were targeted for genetic screening, each for what were viewed as potentially health-threatening conditions. Both groups suffered consequent social injuries.

In a selected body of research, XYY males were linked to criminal or violent behavior.[21] These studies, however, were highly controversial, and their methodology was seriously questioned.[22] It was certainly apparent that most XYY males were not involved in deviant behavior. Through prenatal screening the extra Y chromosome could be detected, and the question arose as to whether parents should abort such a fetus.[23]

Two significant dilemmas emerged over these XYY detections. The first was the inevitable anguish for parents who had to determine whether an XYY fetus was in fact "diseased." Should they consider abortion on this basis? It must be recognized that given the scarcity of reliable information on this subject, such a deliberation could operate only in a state of extreme uncertainty. If the parents chose not to abort, a second problem emerged. Given their knowledge, would this heightened sensitivity induce parents to evoke the very behavior they were seeking to avoid? Would parents be on constant alert for manifestations of the syndrome? Would the child's life inevitably be molded by the known fact of his XYY status? Barbara Katz Rothman reports of one parent's dilemma: "It's hard enough to raise a normal kid. If he throws the blocks across the room will I think he's doing it because he's two, or because he's XYY?"[24]

African Americans represent another group at labeling risk in recent decades. As a group, they carry the sickle-cell trait more frequently than the rest of the population. Although carriers do not exhibit any symptoms of the disease sickle-cell anemia, the couples who both carry the trait have a one-in-four chance of having a child with sickle-cell anemia.[25] In the 1970s serious concerns over the spread of sickle-cell anemia led to extensive pressure to have all African Americans tested, both potential parents and neonates. It should be noted, however, that sickle-cell anemia also exists among Mediterraneans and other ethnic groups, although at lower rates.

Although perhaps well intended, the plans for genetic screening of the sickle-cell disease were highly stigmatizing. One New York State law, for example, stated "such test as may be necessary shall be given to each applicant for a marriage licence who is not of the Caucasian, Indian or Oriental race, for the purposes of discovering the existence of sickle-cell anemia."[26] Of this law, John Fletcher noted, "This extraordinary statement obscures the fact that others than blacks may carry the trait and have sickle-cell anemia. The omission speaks loudly of racial obfuscation." Similarly Fletcher notes that a District of Columbia law referred to sickle-cell anemia as a "communicable disease," yet "the term 'communicable' in ordinary language means something quite different from an inherited genetic condition."[27]

In the early 1970s the sickle-cell condition could not be identified prenatally through amniocentesis. Thus there were no preventive approaches for two carriers of the disease other than to refrain from bearing children. In this context, perhaps the most stigmatizing of all remedies was suggested some years ago by scientist Linus Pauling. In his desire to prevent two carriers from producing children with the sickle-cell trait, Pauling recommended that "there should be tattooed on the forehead of every young person, a symbol showing possession of the sickle-cell gene" so that "two young people carrying the same seriously defective gene in single dose would recognize this situation at first sight, and would refrain from falling in love with one another."[28]

Although the foregoing proposal is almost too absurd to discuss, were it not for the rather famous scientist who proposed it, similar but less extreme proposals were also alarming. In a conference on these issues, for example Evelyn W. Jemison expressed this caution: "I am very much concerned about the psychological impact of sickle cell information on non-blacks that has produced increasing stigmatization of the black school children in Virginia. Why is it that the heterogeneity of the United States population does not result in the testing of all subpopulations, as in the case of PKU testing? Is this in any way a prejudicial bias?"[29]

It is now clear that many States that thought they were instituting *disease*-screening programs were in fact screening for *carrier* status. Because of this confusion, some of the State laws, fortunately, went unfunded or unenforced. Nevertheless, as Tabitha Powledge has noted, "The fact that such testing could be so quickly mandated . . . given the lack of sound medical rationale for such urgent compulsory programs, is an interesting and perhaps disturbing sidelight on the confused interaction between the medical and political processes in the United States."[30]

Perhaps even more important, sickle-cell dilemmas point to the willingness of governments to fund certain types of programs—those with a genetic orientation—while ignoring more critical and broadbased social needs of black Americans. As Dorothy Wilkinson writes, "there are negative political implications to proliferating laws targeted solely at sickle cell with little or no legislation or enforcement of existing laws dealing with the myriad of socio-economic problems confronting black Americans. These neglected areas are also related to their health and welfare."[31]

These labeling dilemmas have not diminished over time, nor are they limited to the United States. In 1987, for example, Australia instituted a mass screening program for the "Fragile X Syndrome." This is a condition in which the X chromosome is likely to break and is a "common cause of mental retardation, second only to Down Syndrome."[32] Once again, concerns over the accuracy and reliability of the test have arisen, and again it is carriers as well as those with symptoms who are being tested. As in the case of sickle cell, proponents of testing argue that carriers "may not know they are affected, but run a high risk of transmitting the condition to their children." As in the previous cases, there are serious dangers of labeling. The Hastings Center, for example, asks, "Will schools require every child who is having trouble with math to undergo a screening?"[33]

Other current dilemmas have emerged over recombinant DNA techniques. Like genetic screening techniques, new recombinant DNA methods have the capacity to define, label, and stigmatize. Some of the newest technologies involve genome mapping which can "define the 'normal' (that is most prevalent) genotype with great precision."[34] This process can indicate genetic markers or proclivities. Yet because most diseases have multiple causes, the exact combinations needed for identification are unknown. Prognosis is uncertain, but identification, labels, and fears are nevertheless created.

Overall, these practices have begun to set a tone of increasing stigmatization. A new array of labels has been created—carrier, XYY, sickle cell, fragile X, genetic markers; others are emerging. At a time when social policies are attempting to diminish the negative consequence of labeling, the evolution of new and powerful labels represents a distinct threat to the social progress so recently achieved. This situation has caused serious problems for adults but has been even more difficult in the determination of "fitness" for newborns.

THE SEARCH FOR PERFECTION

> *Lord Francis Crick, the great Nobel Prize winner in genetics, has proposed that society may have to consider seriously that no neonate should be declared legally human until it is a few days old and has passed a genetic test. Disposing of an infant which fails the test would be different from killing a human being.*
>
> —MILTON HIMMELFARB, "Biomedical Ethics and the Shadow of Nazism."[35]

> *Do we want a world of "perfect people"? I really wonder what are the human costs of attempts to control our differences, our vulnerability. I believe that if women are to maintain our "choice" we must include the choice to have a disabled child.*
>
> —MARSHA SAXTON,
> "Born and Unborn: The Implications of Reproductive Technologies for People with Disabilities"[36]

Our preoccupation with genetic "fitness" has begun to have serious consequences for the anticipatory state of prospective parents. They are now so armed with statistics on the various probabilities of genetic diseases that they are experiencing the childbearing process with acute anxiety. This problem has undoubtedly always existed to some extent but appears to be increasing in intensity.[37]

Pressures to undergo genetic screening and testing are rising. Bearing a so-called genetic failure is coming to be perceived as an act of personal negligence rather than an Act of God. At the same time, parents may take even greater pride in bringing forth a more perfect child. This pursuit of perfection in childbearing is now widely acknowledged. Leon Kass, a bioethicist and social philosopher, has observed that "a new image of human procreation has been conceived and a new 'scientific' obstetrics will usher it into existence. As one obstetrician put it: 'the business of obstetrics is to produce *optimum* babies.' "[38]

The desire to optimize the quality of infants is one of the most significant trends in contemporary obstetrical medicine. With it has come a simultaneous effort to eliminate the less than perfect child. As Marc Lappe has noted, "the allure of a genetic test for a normal or in the future an optimal baby threatens to reinforce an inexorable trend in Western society toward typecasting the less than optimal into categories for assortment and ultimate disposal."[39]

Similarly, Daniel Callahan has argued: "behind the human horror at genetic defectiveness lurks, one must suppose, an image of the perfect human being. The very language of 'defect,' 'abnormality,' 'disease,' and 'risk' presupposes such an image, a kind of prototype of perfection."[40] Through progress in genetic screening, Callahan points out, this prototype has emerged with renewed vigor in recent years.

John Fletcher, in the mid-1970s, specifically noted the potential of genetic screening to undermine increasingly progressive attitudes toward the treatment of infants with disabilities. Fletcher outlined what he termed a historically developing "modern consciousness" concerning disabled newborns which included: an increasingly complex understanding and acceptance of the child; a willingness to attempt to overcome the disability and treat the child as an equal; and an alliance between hospital, health professionals, and parents to maintain the child in a noninstitutional setting wherever possible.[41]

Fletcher argued that the increasing capacity to predict disability threatens to sabotage this modern consciousness in several ways. Among its effects, Fletcher noted an increasing likelihood of *rejecting* the child, because the child "might have" or ("should have") been prevented, an increasing likelihood of *abandoning* the child because of the experience of guilt at having failed to prevent the child, and finally, an increasing acceptance of *euthanasia or infanticide* in that had the child been tested only several months before, it might have been "prevented." Fletcher concluded, "It is reasonable to expect attitudes toward newborn defective children to be affected more negatively than positively by amniocentesis."[42]

On the question of euthanasia, Fletcher's predictions were essentially correct. Whereas several years ago the frequency of this practice was seriously limited, it has now become more common, or at least more highly publicized. Numerous cases of "nontreatment" have been adjudicated through the courts, and the various approaches—withholding surgery and food primarily—have received some degree of legitimacy albeit in the midst of the controversy.[43]

The Baby Doe and Baby Jane Doe cases are the most prominent examples of the euthanasia issue. Through a variety of court proceedings, it is now in the legitimate purview of a family, in consultation with physicians and the courts, to withhold treatment or food. These new rights, for parents and physicians, have been questioned by two groups, not typically in coalition with each other. One group, opposing the rights of euthanasia, have been Right to Life advocates, backed by the Reagan and Bush Administrations. Most of their efforts to create strict medical regulations (including "Baby Doe Squads" and "hotlines" to report nontreatment) were rebuffed by the courts and by most physicians as well.[44]

A second interest group opposing developments in euthanasia has been disability rights advocates, represented by such writers as Adrienne Asch, Michelle Fine, Mary Johnson, Marsha Saxton, and Anne Finger, among others.[45] Their reasoning is particularly relevant to the problem of stigma. They argue that the rising legitimacy of "doing away with" severely disabled infants violates the very basis of the disability movement itself. An emphasis on rights includes the right for a disabled child to live and to receive medical treatment.[46] Thus many disability advocates see the issue not as euthanasia but as infanticide. They cite an unwillingness on the part of parents and society to bear the cost of a disabled child's life and an unwillingness to see a disabled child as fully human. Although there is no unanimity among disability activists, the frequent message of their work is that they would prefer to see greater monies for medical treatments and social services, not 'eugenic intervention.'[47]

Disability rights groups have also expressed concerns over selective abortion on the same grounds that they oppose euthanasia. Their attitudes are much the same as those of women's groups, who regard abortion solely for purposes of sex selection as, at best, questionable. They argue there is an equal disparagement of rights and self-esteem when aborting a child known to have a disability as when aborting a child known to be female. The contention is not over the legitimacy of abortion, which is typically affirmed, but rather over the decision not to have a *particular* child, on the basis of discriminatory criteria.[48] Further, selective abortion of fetuses with "targeted" diseases, such as Down's syndrome and neural tube defects, ignores the high degree of uncertainty in contemporary methods of prognosis. Severity of disease is unpredictable, yet diagnoses are often treated in a wholesale manner, as if all outcomes were alike. In this way, the practice of selective abortion fails to acknowledge the uniqueness of individual responses to disease and disability as well as expectations for progress in social policy and civil rights.

It must be recognized that on the face of it, a prospective parent's desire to have an "optimum" child is an obviously reasonable goal. All parents wish their children to be endowed with as many gifts as possible and spared an equal amount of suffering. Family by family, each would wish for the healthiest possible child. Yet, at a societal level, the steady pursuit of optimization causes a dangerous mind-set. The contrast between what society considers the "better" child and the "worse" child grows more stark. Discrimination seeps into the process, not intentionally but perhaps inevitably.

The essential message of Erving Goffman's book, *Stigma*, is that *all of us* carry an array of qualities that society may stigmatize. These may be physical disabilities, or they may be family disgraces, personal eccentricities, perversities, special talents, or a lack of talent. Whatever their source, social groups will define certain characteristics of human beings as "discreditable." To understand this, then, is to share a sense of identification and empathy with the experience of others.

Yet the tilt of Genetic Welfare creates the reverse perspective. In seeking to optimize the physical entity, it simultaneously creates the vision that disabilities and discrediting attributes are unnecessary and dispensable; that if we properly anticipate them, and act in accordance with knowable biological principles, we may not have to endure their occurrence at all. It thus poses the need for social supports as less necessary and perhaps wasteful "burdens" in a world where biology can be manipulated to prevent those burdens.

If the very groups that are now being assisted through social policies are those that can be eliminated, at or before birth, there will be little societal sanction for social programs. We may slowly shift social priorities away from improving the environment and toward a genetic, not social, vision of human welfare. Policy advances, made in recent decades, may be undermined. As we move toward biological prevention, we may fundamentally alter the perception of our responsibilities toward those seeking redress from societal discrimination.

Finally, it should be noted that the pursuit of human perfection poses a problem beyond the specific individuals and groups involved. It also raises the possibility that the *general* nature of societal tolerance will be altered, our tolerance for all degrees of differentness. The theory of Emile Durkheim, the classic social theorist, provides a useful metaphor for addressing this question.

GENETIC WELFARE AND SOCIETAL TOLERANCE

> *Imagine a society of saints, a perfect cloister of exemplary individuals. Crimes, properly so called, will there be unknown; but faults which appear venial to the layman will create there the same scandal that the ordinary offense does in ordinary consciousness. If, then, this society has the power to judge and punish, it will define these acts as criminal and will treat them as such.*
>
> —EMILE DURKHEIM,
> The Rules of Sociological Method[49]

In *The Rules of Sociological Method*, Emile Durkheim presents a theoretical approach to the nature of social deviance. He addresses such questions as the inevitability of crime, its appropriate treatment, and most important for our discussion, the social tolerance of deviant behavior.

For illustrative purposes, we may think of Durkheim's vision of social tolerance as a continuum. Groups at one pole of the continuum are highly accepting of a wide range of behaviors; groups at the other pole are intolerant of even slight variations in normative behavior. Thus Durkheim's "society of saints," cited above, represents one of the least tolerant points on the scale, where even minor deviations are viewed with alarm and horror. Alternatively, we might say that the social and sexual revolutions of the 1960s made most Western societies broadly tolerant of a wide range of social behaviors, placing them at the other pole of the continuum.

According to Durkheim, tolerance of deviation, in a given society, at a given time, is of a piece. When a society grows more accepting of certain practices it also tends to grow more tolerant of all other behaviors. It is unlikely that a group can free one aspect of deviation from harsh enforcement without simultaneously expanding its tolerance for all forms of deviation. Alternatively, when one becomes less tolerant of certain forms of deviation, that atmosphere of blame also becomes broadly diffused. One might consider, for example, the United States under McCarthyism.

Durkheim argued that "strong states of the common consciousness cannot be . . . reinforced without reinforcing at the same time, the more feeble states."[50] In our effort to reinforce more serious norms, lesser issues of taste and individuality such as dress, speech and socializing must also bear the brunt of official sanction, much as they do in orthodox, traditional, or highly disciplined societies such as the Amish, Hasidic Jews, the military, or contemporary Iran.

If we should attempt to eliminate extreme forms of deviation, according to Durkheim, we would not do so at all, but would instead merely shift the objects of societal scorn to lesser issues. In the effort to wipe out major deviance, we would simply become more intolerant of small violations. This being the case, Durkheim argued, deviance is inevitable; it is bound up with the fundamental conditions of all social life. To attempt to eliminate crime and deviance, Durkheim argued, is futile, for such an effort results merely in a redefinition of deviance, not its elimination. We would simply arrive, Durkheim argued, in the same situation as "the perfect and upright man" who "judges his smallest failings with a severity that the majority reserve for acts more truly in the nature of an offense."[51]

Finally, to Durkheim, deviance is functional. It permits variety and diversity, fending off the dangers of homogenization. From Durkheim's perspective, deviance should be *valued,* for it promotes social change; it symbolizes an individual's ability to act and think differently from others. Socrates, Durkheim notes, was a criminal under Athenian Law. Other historical examples of criminals and deviants might include Galileo, a heretic under religious law, and, as a contemporary example, Martin Luther King, a man imprisoned for defying the laws of his time. As Durkheim observed, in one of his most interesting conclusions, "In order that the originality of the idealist, whose dreams transcend his century may find expression, it is necessary that the originality of the criminal, who is below his time, shall also be possible. One does not occur without the other."[52]

Durkheim's framework suggests one possible role that differentness, in its many guises, performs at a societal level: it promotes tolerance. Given the existence of a broad enough spectrum of differences, we come to accept or even cherish the lesser differences of behavior and type. In turn, we become a society more capable of innovation and change. In the metropolis, for example, unlike the small town, the level of diversity—and hence social change—is great. Indeed it is precisely for these characteristics of diversity, tolerance, and innovation that millions have migrated to cities throughout history.

If we now attempt to apply Durkheim's framework to biological differences, the issues may be much the same. Theoretically, if we seek to "weed out" the more obvious defects of genotype and phenotype, might we not ultimately make more significant and become more sensitive to lesser differences of form and type? As major differences have come to be grounds for biological elimination, might not smaller differences later come to be equally legitimate targets?

From this analogous perspective, it might be concluded that little progress may ultimately be made in an effort to extirpate biological imperfections. As Durkheim's "society of saints" and his "perfect and upright man" viewed their imperfections, so we might view ours. To take the argument to its extreme, we might eventually perceive, with alarm and abhorrence, short necks, skewed noses, weak chins, receding hairlines, or a host of other imperfections. Equally likely would be a heightened or renewed attribution of deviance to distinctive racial and ethnic characteristics.

These speculations raise an important set of questions. If, as Durkheim argues, we must permit extreme forms of deviation in order that individuality may thrive, must we also "permit" extreme genetic disease so that simpler forms of diversity may be allowed? For example, is the existence of mental retardation necessary to the permissibility of eccentric genius? Is a society that permits a child to live, who has neither arms nor legs, necessary to allow a Toulouse Lautrec to walk amid his society with merely shortened legs. Must the more serious forms of illness exist so that we do not "eliminate" those who have epilepsy such as Dostoyevski. In short, do the demands on societal tolerance to accept the most difficult forms of deviation permit us to accept or allow the lesser differences of nature?

If, for the sake of argument, we accept Durkheim's analysis and its biological applications, we are faced with an assessment that is, at best, enigmatic. The analysis suggests that an accepted societal good—the prevention and elimination of disease—may have serious costs or latent consequences for the society at large. It is as if to say that the existence of biological disease, perhaps tragic for the individual and his or her family, may represent a societal strength in its ultimate consequences and serves a functional purpose. Inversely, and perhaps more logically, it argues that an intensive attempt to eliminate biological differentness would be seriously threatening to the social order. It could impair the fundamental tolerance level of society as a whole.

There is, of course, what is often called a "functional trap" in the preceding analysis. For example, many writers have taken all varieties of undesirable social conditions, such as inequality, social stratification, poverty, etc., and shown that they too are functional; they serve a purpose in the larger scheme of things. This then becomes a reason to preserve the status quo and fend off social change. "Things are best left as they are" is often the ultimate resolution of such an approach. It is clearly not the intent of this analysis to suggest that a status quo of suffering and disease is desirable and should be defended. Rather, the present argument seeks to assess the severity of social change that might emerge from Genetic Welfare, for such consequences will have to be seriously addressed in the light of new technological developments.

In further opposition to the Durkheimian argument presented, it might also be argued that genetic difference need not be the only avenue through which a society can reinforce its norms of tolerance. There will, of course, always be groups that warrant social concern—the poor, those disabled later in life, the elderly, war victims, and so on. But it must also be noted that if Durkheim's argument is essentially correct and tolerance is of a piece, then if one group—say, those born with genetic disabilities—becomes so stigmatized that we systematically seek to prevent its appearance, it is possible that other groups will suffer increasingly harsh treatment as well. In short, if we are unable to accept certain types of individuals at birth, it is unlikely we will be more tolerant of those whose characteristics appear at a later date.

The foregoing analysis is theoretical and frankly speculative. But it does suggest that a special class of problems may emerge from a wholesale effort to root out deviance and disability. It argues, with Durkheim, that we may become less accepting of more minor conditions if we eliminate severe variation. Finally, it should be noted that there is, in fact, some evidence to indicate that these potential effects on disability may already be happening. For example, the question has arisen as to whether to abort not only fetuses with highly disabling conditions, but those with more moderate conditions such as diabetes and hemophilia. Both of these illnesses are treatable, and while lives may be difficult, they are not usually catastrophic impediments to a full life. Such a move, if pursued, might very well represent a further step along the continuum of intolerance.

CONCLUSIONS

The worldview of Genetic Welfare has the potential to undermine recent social advances in the treatment of deviance and disability. The programs now achieved are ones that enhance the acceptance, involvement, and civil rights of those who are "different." Policies such as mainstreaming and deinstitutionalization communicate messages of tolerance. They say "You are welcome in the social community; if arrangements need to be made to accommodate you, we will do so." Genetic Welfare, in its extreme form, may communicate the reverse message. It says, "we are deeply afraid of your differentness and will do a great deal to prevent your appearance."

On reflection, there is a certain ferocity about a single-minded effort to extirpate deviance and disability. It is, in effect, an avowal of intolerance. It states, "we will not permit you to pass through out gates"; "we do not want you with such intensity that we will invest significant sums of money to prevent your appearance"; "we do not want, or refuse to be, burdened with the costs your life will incur." It is somewhat reminiscent of the "old George Wallace," swearing before his constituency, "I say segregation now, segregation tomorrow, segregation forever."

Moreover, the problem of segregating—denoting, defining, and labeling—appears to be increasing with developments in genetics. At the very time when social movements are seeking to avoid the stigmatizing effects of labels, new labels are emerging, with a new genetic content. These new labels threaten to undermine recent social gains in public policy. Their potential social consequences are significant.

Finally, in regard to the question of social tolerance, it would appear that a new consideration of genetic intervention is necessary. While the treatment and prevention of illness are obviously positive and worthwhile endeavors, the *vigilance* with which we pursue these goals may have significant latent consequences for our ability to care for and about those society deems "less than perfect." Ultimately we may jeopardize the nature of social tolerance as a whole. Eventually, we must temper the pursuit of optimization with acceptance for those who do not meet rising standards of genetic perfection. To achieve that balance will be a fundamental challenge for the future.

4

THE PROBLEM OF POWERLESSNESS

Civilization's going to pieces. . . . Have you read "The Rise of the Colored Empires" by this man Goddard? . . . Well, it's a fine book and everyone ought to read it. The idea is if we don't look out the white race will be—will be utterly submerged. It's all scientific stuff; it's been proved. . . . This fellow has worked out the whole thing. It's up to us who are the dominant race, to watch out or these other races will have control of things.

—F. SCOTT FITZGERALD,
The Great Gatsby

The great problem of civilization is to secure a relative increase of the valuable as compared with the less valuable or noxious elements of the population. . . . The problem cannot be met unless we give full consideration to the problem of heredity. . . . I wish very much that the wrong people could be prevented entirely from breeding. . . . The emphasis should be laid on getting desirable people to breed.

—THEODORE ROOSEVELT

The interaction between science, politics, and the public has systematically generated an interest in elites, superior men and women, and the desire to perpetuate certain types of human beings. These are issues of power and privilege, influence and dominance.

Since at least the time of Plato, there has been a recurrent desire to propagate a higher level of human being.[1] This desire has periodically manifested itself in social movements that have sought to favor certain groups for reproduction while discouraging or preventing others.

Given this historical context, several questions must be posed. To what degree does such elitism permeate the contemporary worldview of Genetic Welfare? What is the likelihood that genetic technology will affect the social stratification of present and future societies? Will the conflicts between the "haves" and "have-nots," of current and projected societies, be heightened or eased? These questions constitute the central issues of this chapter.

POWER AND SOCIAL CHANGE

Throughout history there has been much speculation and great disagreement about the nature and source of power. Contemporary debate has centered on such questions as who holds power in a given society, whether power is differentially distributed, and how power is exercised. Consensus on such questions is rare.

For purposes of this discussion, several assumptions will be made. First, that power is pyramidic: that at any given time and place, a few groups wield significantly greater power than others. Second, that those in power and those who are powerless are often identifiable as groups, with distinctive social identities. Such groups are typically class based, but may also be sex based, racially, religiously, or ethnically based. Over time, groups may change in their relative strength; a few may become more broadly empowered while others may become oppressed. Many groups remain at the bottom of the social structure for centuries.

These assumptions reflect a "power elite" perspective advanced by such authors as C. Wright Mills, Floyd Hunter, G. William Domhoff, Frances Fox Piven, and Richard Cloward.[2]

This theoretical orientation sees the "power structure" as entrenched. To gain power, "out-groups" must protest or organize, challenging the system for even small shares of social power and social equality.

In recent decades, protest movements seeking increased power have been commonplace. In the 1950s, the civil rights movement worked to empower minority groups; the War on Poverty of the 1960s similarly addressed the "powerlessness of the poor." The women's movement beginning in the 1970s and the more recent disability and gay rights movements have also sought to redress inequities in power and influence. The actions of these groups have not fundamentally altered existing power arrangements, but some gains have been made. Voting, educational, and occupational opportunities have expanded. Affirmative action has addressed some grievances from the past.

These advances have been long in coming and are mercurial at best. The Reagan and Bush administrations, for example, have severely weakened affirmative action. "Last hired, first fired" is still the predominant response to economic recession, seriously affecting disadvantaged groups. More subtle forms of de facto bias are rapidly replacing Jim Crow and de jure racial segregation; the same could be said of women's rights, those of the poor and the disabled.

Traditional power positions are not easily shifted. Gains made must be nurtured, held together by a strong ideology of social change. When empowerment and enfranchisement are moving forces in society, as they were in past decades, progress may be made. When patterns of exclusion are justified, gains may be reversed.

Technology is one element that may alter the direction of such social change. If a new technology expands the pool of available resources, it may create paths of social mobility for those at the bottom of the social hierarchy. Alternatively, if a technology gravitates to the powerful, it reinforces the status quo. It may disarm arguments in favor of new groups and set back progress. In the following sections, the contemporary tilt of genetic welfare will be considered in relation to issues of power.

GENETIC WELFARE AND ELITISM

Elitism, in a genetic context, may be thought of as the efforts of the powerful to promote the fortunes of the superior, gifted, or more "fit." It is a means of maintaining the power distribution of a society. To favor those already advantaged and to further the welfare of the "fit" may at minimum, solidify the power positions of a select few, and at worse, actively seek the elimination of those deemed unworthy.

The joint precedents of eugenics and Nazism are, of course, the most dramatic historical examples of elitist genetic visions. Clearly, no scientist is today proposing a "master race," but the very hint of biological selection understandably raises concerns. As an introductory caveat, it is important to note that overt patterns of elitism are often aberrations of contemporary genetic concerns and do not typify mainstream thought. Yet it is precisely such deviations that tend to contaminate other endeavors and which thus need to be assessed. They may be indicants of future issues and signposts of coming trends.

One of the earliest examples of modern genetic elitism began during the Nazi era. In the 1930s Hermann J. Muller, an American geneticist and winner of the Nobel Prize, advocated a program of "eutelegenesis," that is, the selection of "sperm from superior men" for purposes of widespread artificial insemination.[3] In the 1960s Muller continued to urge a program of "germ-cell choice."[4] He prescribed selecting for artificial insemination "people who are decidedly superior in endowment," and "genetic material from outstanding sources."[5] Muller clearly recognized the social implications of his plan, labeling it a "Genetic Operation Bootstrap" which should ultimately "be incorporated into our mores."[6] Ideally, for Muller, this practice would be institutionalized as a basic building block of human reproduction.

Muller's proposal exhibits several dangers of genetic elitism. One of the most dramatic problems was Muller's changing definition of "superior men." Early in his career, Muller was intrigued by the political left and proposed, as models for superior genetic perpetuation, figures such as Marx and Lenin. By the 1960s, Muller's heroes had changed. While still promoting eutelegenesis, Muller wished to replace Marx and Lenin with more traditional leaders such as Einstein, Pasteur, Descartes, and Da Vinci.[7] The implications are clear. Changing political fortunes and shifting societal definitions alter "heroic stature." Greatness is often bracketed by time and place.

Had Muller's plan been pursued, as he desired, the early offspring would have been soon out of fashion. The capacity of "select" future individuals on the basis of existing political and social reputations is a clearly perilous path. This dilemma, of course, only compounds the seriously questionable assumption that positive leadership traits are inheritable, a perspective for which there is little evidence of merit.

This quest for "superior types" has not dissipated. In the 1980s the search reemerged in the form of the "Shockley controversy." In an attempt to implement Muller's program of eutelegenesis, a California sperm bank sought to acquire the sperm of Nobel Prize winners, for purposes of artificial insemination. William Shockley, a Nobel Prize recipient, for the invention of the transistor, publicly acknowledged his several personal "donations" to the clinic. Four other Nobelists, unnamed, also were said to have contributed.[8] The clinic's goals in the creation of "superior individuals" and the participation of figures such as Shockley suggest that the elitist emphasis of classical eugenics has not yet vanished. The continuing belief that superior intelligence can be genetically endowed remains a powerful force in contemporary society.

The perpetuation of superior types is not the only elite bias of genetic and reproductive engineering; elitism also manifests itself in the clientele to whom technologies are available. "Genetic manipulation is a piece of luxury technology," according to Kirsten Aner.[9] The understanding of its potential, and access to its benefits, belong primarily to those of the upper and middle classes; "knowledge, as always, gives power; and also, as usual to those that have, more is given. All experience shows that genetic counseling and screening always interest a sophisticated and well-endowed *elite* of the people."[10]

The technologies of in vitro fertilization, surrogate parenting, artificial insemination, genetic screening, and genetic surgery, to name but a few, are expensive. Typically costing in the thousands of dollars and sometimes tens or hundreds of thousands of dollars, their benefits are rarely covered by insurance, even less by Medicaid. A few technologies, such as the freezing of embryos, have cost more than a million dollars.[11] Benefits, therefore, are primarily available only to those who have monies for luxury items, to those who can afford elective procedures in which payments are often made in large lump sums. This represents a de facto limitation of these technologies to the well to do.

This unequal distribution of reproductive technologies will likely result in distinctions of rights and privileges. The lower classes will not have choices even in this arena, choices which may be available to other groups. Perhaps even more important, the unequal distribution of specifically medical techniques, such as genetic screening, counseling, and surgery, will likely compound already existing health differentials in our population. If those already favored have opportunities to avoid genetic diseases, while those less well off are rarely able to have such opportunities, the consequent medical and social drain will further the disadvantages of the underclasses.

Equally important, investments in expensive technologies divert attention from social efforts that offer specific benefits for the less advantaged. A significant example lies in genetic screening for mental retardation. Trisomy 21, or Down's syndrome, is of course widely dispersed along the social hierarchy. Yet scientific evidence, accumulated over several decades, consistently demonstrates that genetic impairments account for most upper- and middle-class retardation, but are the source of much less retardation among the lower classes.[12]

In the lower classes, retardation is often a result of poor prenatal care, denial of sensory stimulation, lead poisoning, or a failure of the schools. In short, it is a *social* product. But the increasing use of genetic screening and counseling emphasizes only genetic fallibility and therefore highlights only a portion of the problem. If we continue to address retardation solely as a genetic aberration, we will systematically ignore the thousands who suffer the social, not genetic, disability.

Genetic screening, counseling, and selective abortion unquestionably serve to diminish the prevalence of genetic retardation; the available methods are relatively effective strategies against the presence of the disorder. The clarity of the physical defect makes the genetic aspect of the problem fundamentally understandable. Environmental retardation, however, is less direct in its causes and certainly less immediately addressed.

The genetic problem requires the cooperation of relatively few individuals—doctor, counselor, family—not many others. The solution to environmental retardation requires the cooperation of multiple and competing interests, extensive institutional change, and a recognition of great social failings. Logically, society and medicine will pursue what can be done efficiently, effectively, or profitably; the genetic emphasis in such a choice will undoubtedly prevail. Thus the problem of mental disability at the lower class levels will continue to receive far less attention. As we elect to focus vast financial and scientific resources on genetic rather than social retardation, advantages once again accumulate to those of the middle and upper classes, at the expense of the now disadvantaged.

A final consideration of genetic elitism involves surrogate mothering. The formal contractual arrangements of surrogate parenting imply a quid pro quo of money in exchange for birthing. It is in the breakdown of these contractual agreements, however, that problems of stratification and power become conspicuous. The Mary Beth Whitehead dilemma is a case in point. Recruited to serve as a surrogate mother, Mary Beth Whitehead changed her mind. Her wish to keep her daughter led to a controversial court case in 1987. In the testimony that determined custody of the child—to the Whiteheads or to the father's family, the Sterns—the battle was waged as much in terms of social influence as it was in terms of contract law.

Whitehead and her family were consistently pictured as of the working class—unstable, emotional, and without resources. The Sterns, upper middle class, one a doctor, the other a biochemist, were portrayed as resplendent in the benefits they could offer the child; their intellectual, financial, and social advantages were heavily stressed. As Rita Arditti notes, "Judge Sorkow came down hard on Mary Beth Whitehead characterizing her as 'manipulative, impulsive, and exploitive,' while he praised the 'stable environment of the Sterns.' He described the Sterns as 'credible, sincere, and truthful people who would initiate and encourage intellectual curiosity for the child.'"[13] Despite the intricacies of the legal outcome and appeals, it should be noted that both the media and the original settlement suggested that the enhanced social standing of the Sterns gave them greater *entitlement* to the child. The vision of justice that initially prevailed was one of benefit to those already empowered. It is important to consider whether, in the continuing determination of such cases, a similar tilt will occur.

In sum, if we examine the accrual of benefits, it would appear that genetic technologies have the distinct capacity to enhance the social, medical, and intellectual advantages of the already powerful and prosperous. They are of considerably less benefit and possible harm to those at the bottom of the social hierarchy. While these inequalities may not significantly alter the status quo, they may certainly function to maintain it.

Bioethicist Marc Lappe has argued that all eugenic influences are dangerous for they reinforce notions that "the present division of wealth and power corresponds to some deeper reality of human life," and that one such deeper reality is "genetic difference."[14] Yet the desire to justify inequalities has always been a powerful historical force. Edward Bellamy, as but one example, writing in nineteenth-century America, metaphorically posed the status hierarchy of his time as a carriage: those sitting on the top having the advantages, those less fortunate, walking beside or pulling the carriage. He too noted that those who arrive at the top of the carriage come to share what he called a "singular hallucination," that their favored position is fully earned and correctly reflects their superior effort and worth. They believed "that they were not exactly like their brothers and sisters . . . but of some finer clay, in some ways belonging to a higher order of beings. . . . The conviction they cherished of the essential difference between their sort of humanity and the common article was absolute."[15] A rationale for unequal status has existed in virtually all societies and in virtually all epochs. What then is different in this new genetic context?

It may be argued that genetic technologies expand and accelerate the process of social differentiation. Indeed their very purpose is to identify and isolate human differences so that medical action may be taken. But, as noted in Chapter 2, medical judgments yield quickly to social judgments. The pattern of access to expensive medical resources helps some groups and harms others. The focus of medical research, such as that on mental retardation, systematically ignores the disadvantaged. The medico-legal judgments of our society forge a vision of rights and responsibilities that enhance the advantages of those who are most powerful. Thus, while the accumulation of advantages to an elite is certainly not unusual in the historical introduction of technologies, the costs and benefits in this case are of extraordinary range, influence, and speed. Further, through genetic manipulation, this existing stratification system may also be projected directly into succeeding generations.

GENETIC PLANNING: A VISION OF THE FUTURE

> *Each generation exercises power over its successors: and each . . . limits the power of its predecessors. . . . if any one age really attains, by eugenic and scientific education, the power to make its descendants what it pleases, all men who live after it are the patients of that power.*
>
> —C. S. LEWIS,
> The Abolition of Man[16]

Who has the power to shape our genetic future? Whose vested interests will prevail? What groups will be included, whose needs will be ignored? These questions consider ways that contemporary Genetic Welfare may mold the stratification systems of future societies. At issue is the perpetuation of existing inequalities.

The quest for genetic improvement has created a new type of futuristic thinking—a genetic planning, if you will—much like contemporary urban or economic planning. While there are, as yet, no five- or ten-year plans, the prospect of designing our descendants entices many with its possibilities for creative artistry. Not only has our current technological situation been viewed as an opportunity, but to some, such planning is perceived as a responsibility. Indeed, it may be argued that if genetic duty becomes the ultimate responsibility of parents, genetic planning may become the ultimate responsibility of society at large.

What are "our obligations to future generations" is the question most typically posed. It is in this context that notions of "engineering" often arise. Martin Golding once referred to this effort as "doctoring" the future.[17] Golding posed his futuristic questions as three:

> (i) Is there now an obligation to bring about any given situation in the future?
> (ii) if there is such an obligation, on whom does it fall; and,
> (iii) if there is such an obligation, what ought to be done now to fulfill it?[18]

Golding's questions engage several dimensions of power: Who decides that such an obligation exists; whose vision of the future will dominate; what group will undertake responsibility for implementation? Most writers have skirted the thornier aspects of implementation and enforcement, but many suggestions for an ideal future species have been advanced.

Hermann J. Muller, once again, was one of the earliest exponents of genetic planning. Like many of those who followed him, he created a compendium of characteristics that might ideally characterize future human beings. For example, at the intellectual level, Muller argued: "it is obvious that tomorrow's world makes desirable not only a lively curiosity but also a much greater capacity for analysis, for quantitative procedures, for integrative operations, and for imaginative creation."[19] On the humanistic side, Muller called for a "strengthening and extension of the tendencies toward kindliness, affection, and fellow feeling in general, especially toward those personally far removed from us."[20] Physically, Muller sought to improve "the genetic foundations of health, vigor, and longevity; reduce the need for sleep, bring the induction of sedation and stimulation under better voluntary control, and to develop increasing physical tolerances and aptitudes in general."[21]

The package is intriguing, yet, of course, it is not the specific attributes that need be questioned. What is of concern, however, are its fundamental assumptions—first, that such broad traits are inheritable and second, that it is desirable to *plan* for such outcomes. Of particular consideration is the identification of such complex traits as kindliness, affection, intelligence, and endurance, as though there were precise genetic paths for their occurrence.

The design of human attributes did not end with Muller. Many contemporary scientists and philosophers have advanced similar suggestions, each incorporating their vision of necessary genetic traits for the future. Joseph Fletcher, for example, once offered the following list:

Minimal intelligence
Self-awareness
Self-control
A Sense of time
A Sense of future
A Sense of the past
The capability to relate to others
Concern for others
Communication
Control of existence
Curiosity
Change and changeability
Balance of rationality and feeling
Idiosyncracy
Neo-cortical functioning[22]

Scientist Robert Sinsheimer similarly advanced the following "distinct characteristics of humanity which—as we seize our destiny, we may wish to enhance":

Our self-awareness
Our perception of past, present, and future
Our capacity for hope, faith, charity, and love
Our enlarged ability to communicate and thereby to create a collective consciousness
Our ability to achieve a rational understanding of nature
Our drive to reduce the role of fate in human affairs
Our vision of man as unfinished[23]

As a last example, here are Alasdair MacIntyre's "Seven Traits for the Future" presented in 1979:

> Ability to live with uncertainty
> Roots in particularity
> Nonmanipulative relations
> Finding a vocation in one's work
> Accepting one's destiny
> Hope
> Willingness to take up arms[24]

MacIntyre's ironic conclusion is revealing:

> If in designing our descendants, we succeeded in designing people who possessed just those traits that I have described, we should have contrived for ourselves descendants who would be unable, by virtue of those very traits, to adopt manipulative bureaucratic modes of planning. What we would have done is to design descendants whose virtues would be such that they would be quite unwilling, in turn, to design their descendants. . . .
>
> It turns out . . . that it would clearly be better never to embark on our project at all. Otherwise we shall risk producing the descendants who will be deeply ungrateful and aghast at the people—ourselves—who brought them into existence.[25]

Thus it may be argued, as MacIntyre does, that genetically planning the future implies a rigidification of the social structure, so unappealing that "self-actualized" human beings would be loathe to impose such a vision on others. Like our views of heroic stature, noted earlier, our vision of future human needs and capacities will undoubtedly change. As Roger Shinn has cautioned, "There is something frightening in the thought that a generation of mankind, notorious for its social problems, should set out to determine genetically, the future, according to its present values."[26] In this context, the fear arises that certain changes, if pursued, could be irreversible; a single generation would have imprinted itself on all succeeding generations, and within that generation, by a very few individuals. This is power of a new and striking variety.

In its most benign form, genetic planning would occur within families. This is an enterprise that has already begun. Yet even this right has been questioned by some. Sumner Twiss, for example, asks "by what authority does the prospective parent have the right to decide who shall live or die; or what kind of person should be permitted to be born? Is it justifiable for parents or society to articulate standards of normalcy . . . on the basis of genetic makeup?"[27]

Even more problematic are proposals for mandatory screening according to specified standards. Although this is a highly unlikely outcome, the possibilities have been debated. Michael Lerner, for example, has argued that the "emergence of a dictatorship can lead to imposition of sterilization and detection programs for certain classes of the population, for example, those adjudged to be aberrant or insane by virtue of either their *descent* or *dissent*."[28] Though this extreme eventuality is remote, Lerner's fear illuminates a fundamental problem in designing the future. Such plans impose a "select" vision of the future; they attribute authority to "select" individuals; and they open avenues for negative sanctioning of the "less than select." A new form of elite power relations is generated within a new sphere of influence—the power to shape the future.

Other considerations for the problem of power are even more distant, such as the technique of cloning. Though the technology is not available for human application, and likely never will be, it constitutes a persistent metaphor for power in its most pure and potent form.

CLONING: A POWER METAPHOR

"If a superior individual (and presumably genotype) is identified, why not copy it directly, rather than suffer all the risks of recombinational disruption, including those of sex."[29] This comment, made by Joshua Lederberg, a Nobel Prize-winning biologist, became one of the most controversial utterances of the biological revolution. The statement was made rather blithely in 1966, before most of the serious bioethical concerns had begun to be debated. Lederberg later retracted his seemingly casual approach, but the clonal metaphor continues to raise questions.[30] The concern over perpetuating "superior individuals" dominates the clonal controversies.

Cloning is a procedure for duplicating an existing organism; as such, it magnifies the problem of power many-fold. Theoretically, it goes beyond the quest for *types* of traits desirable for the future to the notion of precisely *replicating* specific human beings. It has been much debated despite its practical remoteness as a technology.

As a metaphor, cloning permits us to view the process of selectivity in its extreme form.[31] It is a password for the unwillingness to take genetic risks, a yearning to hold on to what we *now* perceive as desirable with little or no faith in future recombinational possibilities. It is the ultimate adhesion to the present or the past. Moreover, cloning is typically envisioned as "serving the social good" and "for the benefit of humanity." As such, it is a very pure form of the worldview of Genetic Welfare.

Joseph Fletcher, writing in the *New England Journal of Medicine*, has suggested an additional and related purpose for cloning—that of specialization. He proposed the following extreme and unlikely scenario. "If the greatest good for the greatest number (i.e. the social good) were served by it, it would be justifiable . . . to specialize the capacities of people by cloning or by constructive genetic engineering. I would vote for cloning top grade soldiers and scientists or supplying them through other genetic means if they were needed."[32] Other writers have suggested not "top-grade" but "lower-grade" specialization, for purposes of doing society's dirty work: a corps of worker-clones echoing *Brave New World*.

It is not such farfetched situations, however, that have engaged the attention of most scientists, but rather the more expected powers that might emerge from the prospect of cloning. Medically related possibilities, for example, include a "bank" for organ and tissue transplants, or ready access to replacement limbs. Robert Edwards, IVF scientist, notes "this technique [cloning] may be utilized eventually for the benefit of humanity in directions which we do not apprehend today, such as averting the rejection of transplanted embryonic cells. Each new discovery should be examined dispassionately and its benefits weighed in the balance."[33] As we examine these possible benefits, we need to consider *who* would use them and who would have the power of selecting those beneficiaries.

Concerns about potentially nondemocratic uses of cloning were forcefully expressed in the 1970s by James Watson, Nobel Prize winner and one of the discoverers of DNA. In a series of articles, Watson predicted the likelihood of human cloning within twenty to fifty years.[34] Watson's perspective was cautionary; his concern was that asexual clonal reproduction should not emerge in society prior to the public's readiness for it. He advocated "wide-ranging discussion" on this matter "far too important to be left solely in the hands of the scientific and medical communities."[35] His concern was that too much was at stake, and too many opportunities would emerge, to leave serious consideration of this technology to just a few.

The concept of clonal reproduction of a full human being violates many explicit and implicit assumptions of the human species. For example, at least theoretically, the process could create one or more exact replications of a particular human being. This would be done in an effort to *save* a particular genotype because of his or her extraordinary contributions or abilities. The process would violate, at minimum, our historical understanding of the right of human uniqueness. Indeed the effort would be done for the diametrically opposite purpose: not to allow a new human being to thrive, but rather to repeat, as precisely as possible, the past performances of another human being. The psychological burdens on such an individual seem, on the whole, appalling. Of equal concern is whether such a being would in fact be viewed as fully human or instead as a hybridized product, laboratory created and obviously experimental in nature.

The issue is admittedly extreme, yet serves to telescope many of the dangers implicit in both reproductive and genetic engineering. The clonal metaphor represents the ultimate selection process: the power not only to mold our descendants but to identify them directly. It encapsulates the notion of "master race" and genetic planning in its final extreme form: the precise duplication of selected human beings.

A final concern regarding cloning, as well as most other futuristic technologies, is that virtually all the hypothetical historical and contemporary candidates are male. For example, through eutelegenesis, Hermann J. Muller sought to reproduce "superior men"; the California Artificial Insemination Program banks sperm of Nobel Prize winners; likewise, cloning candidates have been similarly idealized male figures. Thus, for feminists, these techniques offer, for the future, but another opportunity for sexual discrimination: a new vision of single sex role modeling. Other genetic and reproductive technologies offer similar possibilities. Sex selection and artificial wombs, like cloning, are potentially far-reaching in their scope; they have the ability to alter the very basis of male-female relations, deeply affecting considerations of sexual status and power. At risk is whether women will be able to maintain the small gains in power achieved in the past few decades; will Genetic Welfare resurrect, in new form, the patriarchal paradigm?

SEX AND POWER: GENETIC AND REPRODUCTIVE CONSIDERATIONS

A significant aspect of the women's movement has been a submovement concerned specifically with women's health. Popularly called "the women's health movement," it has focused national and international attention on the status of women's health.[36] Much of the literature documents the frequency of women's victimization by new medical developments. Genetic and reproductive initiatives have been areas of particular concern.

Fertility drugs and contraceptives, for example, have provided important new opportunities for women, but at considerable personal risk. Medical breakthroughs often have been experimental and typically lacking in proper safeguards. There have been tragic costs: Thalidomide, DES, Depo Provera, and the Dalkon

Shield, to name but a few.[37] Overall, women have been unable to monitor the safety of new techniques or their long-term consequences. Now, still other possibilities present themselves for future consideration. Their effects may be less physically dangerous, but of equal impact on women's social standing.

Ectogenesis, or the "artificial womb," for example, offers the possibility of fertilization and maturation outside the woman's body. Although the technique has not yet been achieved, advances in in vitro fertilization and incubation represent important stepping stones in this direction. Theoretically, if in vitro fertilization can be extended forward in time, and incubation can be extended backward, such that a fertilized egg can be kept alive indefinitely in an artificial environment, the technique will be achieved. Developments in both of these related fields are proceeding at a rapid pace; however, the final accomplishment of this technique is both uncertain and clearly futuristic. Nevertheless, like cloning, ectogenesis has become a highly discussed metaphor of rights, privileges, and power.

To some women, the advent of the artificial womb represents the quintessence of female freedom. Loosened from the obligations of fertilization and gestation, women could have children without the physiological "burden." Yet the female birthing function, while sometimes viewed as a historical symbol of oppression, has also been a source of tremendous power. Women were, and are, a necessary component in the survival of the species. Ectogenesis would remove this vital contribution, placing women on an equal par with men: sexual donors, once removed from the long-term consequences.

There may be less freedom here than would first appear. A possible, though perhaps unlikely, scenario is the ultimate denial of the "rights" of fertilization and gestation for women on the grounds of the "best interests" of the child. Such an outcome is not completely farfetched. We may recall, for example, that in the 1940s and 1950s, bottle-fed infant formulas were deemed far healthier and more socially acceptable than breast-feeding. Indeed, in many areas of the United States it was difficult to get a doctor to "permit" breast-feeding. Physicians rejected these requests on the grounds that the mother was deliberately harming the infant. Though this fad fortunately has passed, its occurrence indicates the wrongheaded unanimity that often develops in the medical profession, and how quickly these beliefs can spread to the population at large. Like cloning, the idea of requiring artificial gestation is thoroughly speculative, yet it is worth consideration because of its extraordinary significance.

A more immediate feminist problem is that of sex selection.[38] Amniocentesis and chorionic villus sampling, for example, both report not only a cross section of genetic diseases but, as a by-product, the sex of the fetus as well. Combined with abortion, this then becomes a route to sex selection. Pre-conceptive procedures are also available, although their effectiveness has not been fully demonstrated and they are not widely used. Nevertheless, through genetic screening, sex selection can now be practiced to a greater or lesser degree. Several concerns have arisen.

Internationally, the predominant preference is for male children.[39] Although the tendency is more common in Third World countries, it exists in the West as well. One potential outcome of this preference is an imbalance in the sex ratio, an overabundance of males in the population, relative to females. This already appears to be happening in China and India, and although officially their governments are making efforts to thwart sex selection, the practice still occurs.

In one interpretation, the shift in the sex ratio could make women more valued because of their relative scarcity. If fewer in number, it is argued, women might be seen as special or rare in the human repertoire.[40] More likely, however, would be an internalized vision, by men and women alike, that females are less desirable and less necessary to the society. This would compound, of course, their already traditional second-class citizenship.

The fact that even women have a preference for male children reflects several social patterns. On the most over level, ancient cultural traditions make sons most valued; economic calculations strongly support these views. But even in the West, where such considerations are less prevalent, sons often are preferred. This reflects a common social tendency: that underclasses, or the disempowered of all types, typically internalize negative societal judgments. For example, welfare recipients sometimes condemn others on welfare as being lazy and unwilling to work; the poor tend to see themselves as at least partly blameworthy. The fact that women should view female children as less desirable is not surprising.

The technological capacity to select for sex may enhance the tendency to promote male offspring. Major financing is now being invested in the development of new sex-selection techniques. This suggests that there is a considerable anticipated market for their use. Evidence indicates that the child of choice will overwhelmingly be male.

An additional problem involves birth order. Internationally, families manifest a distinct preference for a male child first, if desiring a female child, then most often as a second- or third-order child. Yet there is much suggestive evidence linking birth order with a variety of behavioral outcomes.[41] Subtle patterns of independence, initiative, and self-reliance are statistically more common in firstborns. Thus the limitation for women, of first-born status, could enhance the very perception of dependency that is already linked to the stereotyped female image.

A final issue in regard to women and power concerns abortion. As noted, abortion has been used as a response to amniocentesis, for the purpose of sex selection. Most writers, even those who support "abortion on demand," regard sex selection through abortion as frivolous at best and torturous at worst. As John Fletcher once noted, being female is not a disease.[42] However, some analysts, including Fletcher, support the right to sex selection on the principle of *Roe v. Wade*: that inquiry into a woman's reasons for abortion are inappropriate. Alternatively, Gertrude Lenzer has argued, "Surely, it can hardly be the legal or moral intent of the Supreme Court's position to guarantee women the right of self-determination for the purpose of discrimination against their own kind by either doing away with fetuses of their own sex or by choosing male children as their first borns by means of newly developed preconceptive technology."[43] Even more emphatically, Tabitha Powledge argues against all forms of sex selection:

> I want to argue we should not choose the sexes of our children because to do so is one of the most stupendously sexist acts in

> which it is possible to engage. It is the original sexist sin. This argument applies to both pre- and post-conception technologies. To destroy an extant fetus for this reason is more morally opprobrious than techniques aimed at conceiving a child of a particular sex, but they are both deeply wrong. They are wrong because they make the most basic judgment about the worth of a human being rest first and foremost on its sex.[44]

If, as is likely, sex-selection technologies advance beyond post-conception identification to pre-conception selection, the abortion issue will be defused, but its implications will not. Problems of sex ratio, sexual devaluation, and birth order will heighten. The functional superfluousness of the natural womb potentially created by ectogenesis could compound this situation. Female status may once again be in jeopardy, ironically, by loss of the very function that historically has been viewed as oppressive.

CONCLUSIONS

Status and power changes are not an inevitable outcome of Genetic Welfare, but they are likely. The creation of new options, in an existing structure of unequal power, suggests that opportunities will accrue to those already favored. Genetic and reproductive engineering techniques are costly; their pursuit requires knowledge and medical sophistication. Their use is most likely to benefit the advantaged, least likely to benefit the underclasses, although their impact will be felt: the gap between the classes, in the receipt of quality medical care, will widen.

If Genetic Welfare continues to lace medical care with an elite bias, the status of the traditionally less powerful must suffer. The disempowered typically improve their status only in eras when *inclusivity* is a dominant social goal, as it was during the years of the civil rights movement. Similarly, progress must falter when *exclusivity* is at a premium. The selective tilt of contemporary genetic thought suggests that the exclusivity outcome is likely.

Furthermore, many reproductive techniques could function not to improve health but to perpetuate images of the more powerful—the successful, the "superior," and the male. They may turn into biological interventions for *selective* ends, not medical treatments, and thus have the potential for extraordinary social consequences. If they are designed not to treat disease but rather to improve the species, they represent a form of genetic planning that echoes "master-race" social thought.

Finally, the more speculative issues of the future—the power to shape our descendants, cloning, and a shift in male-female relations—are matters of concern. Technologies are appearing at a rapid rate, often far more quickly than scientists themselves envision. Huge investments in techniques such as sex selection, in vitro fertilization, and incubation for medical purposes make some of these futuristic procedures reasonable outcomes of current research and development.

It is unlikely that society will embrace these new techniques wholeheartedly, yet their very availability will shape our social thought. If we can promote "superior" human beings, why should we not? If the technique is available, should we not use it? Thus, as argued in Chapter 2, Genetic Welfare will not spring up full blown, but as is more likely, will encroach upon the future, in easy and palatable steps. We may arrive, even before we know we have embarked on the journey.

5

THE PROBLEM OF ALIENATION

after all, the whole work of man seems really to consist in nothing but proving to himself continually that he is a man and not an organstop.

—FYODOR DOSTOYEVSKI,
Notes From the Underground

The doctor spoke dispassionately, almost brutally, with the relish men of science sometimes have for limiting themselves to inessentials, for pruning back their work to the point of sterility.

—EVELYN WAUGH,
Brideshead Revisited

Or, take a surgical operation.
. . . stretched on the table,
You are a piece of furniture in a repair
shop . . .
All there is of you is your body
And the 'you' is withdrawn.

—T. S. ELIOT,
The Cocktail Party

The problem of alienation is pervasive in modern industrial society.[1] The condition is seen at all levels—work, recreation, formal, and informal relations. There is a recurrent sense of separation from ourselves, from others, and from the objects of our work and interest.

In the sphere of medicine, the perception of alienation is particularly rampant. Physicians, nurses, hospital personnel, patients, and social critics increasingly cite the growing impersonality of the medical enterprise. There is a distinct impression that in the training and practice of medicine, human behavior has become more manipulative, more calculating, and less compassionate than in the past.

Some new directions in medicine have begun to address these social problems, particularly in medical schools and hospital ethics committees. Issues of "informed consent" and "truth telling" are examples. Other developments such as "natural childbirth," hospices, and "death with dignity" have arisen from social movements that are also contending with dilemmas of alienation.

The central question of this chapter is whether Genetic Welfare will assist or undermine these recent efforts to address alienation in medicine. Through the eyes of medical practitioners, will we be seen as unique human beings or as objects, things to be manipulated in a laboratory? How commercialized will human beings and their parts become in the development of medical technique? Will medicine emerge with a more humanistic bent, as it once had in the past, or continue to generate a sense that nature is being usurped by mechanical contrivances?

ALIENATION AND SOCIAL DISTANCE

The most important analysis of alienation derives, of course, from Karl Marx. Alienation was to Marx a logical and necessary outcome of an exploitative economic system. When workers are separated from the means of production, Marx argued, the result is a distancing from the very core of existence.

> What then do we mean by the alienation of labor? First, that the work he performs is extraneous to the worker, that is, it is not personal to him, is not part of his nature; therefore he does not fulfill himself in work but actually denies himself; feels miserable rather than content, cannot freely develop his physical and mental

> powers, but instead becomes physically exhausted and mentally debased. Only while not working can the worker be himself; for while at work he experiences himself as a stranger. Therefore only during leisure hours does he feel at home, while at work he feels homeless.[2]

According to Marx, in an exploitative system human beings are estranged from both the objects of their work and the process of production. That which is distinctively human is lost in the experience. Eventually, Marx argued, "As the world of things increases in value, the human world becomes devalued. For labor not only produces commodities; it makes a commodity of the work process itself, as well as of the worker."[3] Ultimately, to Marx, humans are separated not only from the products and processes of their work, but from nature, from other human beings, and finally, from the species itself.

The urban environment is a sphere of modern life that has also inspired many classic discussions of alienation. In the city, social critics have argued, the human being is an object for manipulation and utility. Georg Simmel, for example, wrote that in the metropolis one "is reckoned with like a number, like an element which is itself indifferent."[4] Ferdinand Toennies defined modern urban relations as *gesellschaft:* interactions that are formal, instrumental, and calculating, deriving from the "cash nexus" of urban life.[5] Similarly, Louis Wirth argued that in the "metropolitan worldview" our "acquaintances tend to stand in a relationship of utility to us in the sense that the role which each one plays in our life is overwhelmingly regarded as a means for the achievement of our own ends."[6] The rise of urban, industrialized life has given forth endless visions of calculability, instrumentalism, and alienation.

More contemporary social theorists have observed the same problems as their classical counterparts. Erich Fromm, for example, wrote: "Alienation as we find it in modern society is almost total; it pervades the relationship of man to his work, to the things he consumes, to his fellowman and to himself."[7] Likewise, Charles Taylor, in an unrelentingly critical vein, argued that in modern society, we have "an indefinable sense of loss; a sense that life . . . has become impoverished, that men are somehow 'deracinate and disinherited,' that society and human nature alike have been atomized, and hence mutilated, above all that men have been separated from whatever might give meaning to their work and their lives."[8]

Alienation has been found to exist at all levels of society. Poets, novelists, philosophers, and social critics identify the sense of distance or estrangement as one of the dominating aspects of contemporary experience. Little wonder, then, that medicine too has been overcome by this modern ennui. It is a problem that seriously troubles practitioners and consumers in the field.

ALIENATION AND MEDICINE

The history of medicine has been marked by a steady progression away from the humanistic and organic and toward the technical and specialized. In the process, medicine has become more efficient, more effective, and more alienating. It has moved physicians away from the bedside and toward the machine. The central focus of medicine has shifted from the whole person to the disease. In consequence, the role, attitude, and function of the physician has altered dramatically over the centuries.

In *Medicine and the Reign of Technology*, Stanley Joel Reiser argues that machines and technology now serve as almost impenetrable barriers to meaningful human interaction. Where once doctors observed a patient's mood, considered life-style as a factor in health and disease, and carefully recorded symptoms obtained through lengthy personal testimony, now technology alone dominates the process. Modern technology is rapidly replacing human observation. Indeed, the machine and the laboratory are often more accurate, but a great cost has been paid in heightening alienation. Reiser argues:

> If physicians in general come to accept a fundamentally mechanical view of human beings, in a world that is more and more

> enamored of technology, the prospect for the future of medicine is extremely disquieting. . . . Machines inexorably direct the attention of both doctor and patient to the measurable aspects of illness but away from the "human" factors that are at least equally important. Insofar as technological evidence occupies the time and commands the allegiance of both doctor and patient, it diminishes the possibility that a close personal relationship will develop between the two.[9]

This transition, from a medical orientation in human dimensions to the dominance of mechanization, reflects an evolution that has been occurring for many centuries. In the Western world, the humanistic emphasis in medicine was first introduced by the ancient Greeks. In the Greek worldview the human being was but one element in a holistic context. Human beings and their environment were perceived as one, integrated in a homeostatic balance.

The four humors theory of Hippocrates created the theoretical basis for the Greek vision of harmony. If the human being was not synchronized with its environment, it was thought, humors, or bodily fluids, would accumulate and the person would become ill. On this basis, the Greeks counseled moderation and temperance as basic treatments. The role of the physician was to ensure this balance. A close personal interaction between patient and physician was therefore necessary, guided by strong ethical precepts.[10] This medical worldview was fundamentally superseded first by Cartesian metaphysics and then by germ theory.

In the early 1600s, following the successes of Kepler and Galileo, Descartes argued that the world and all its component parts, including the human body, could be conceived as simple machine entities. This meant that the human being, like any other object, was subject to physical laws. These assumptions facilitated later scientific exploration in medicine, including Harvey's identification of the circulatory system and Leeuwenhoek's work on the microscope.[11] Despite these advances, however, the role of the physician at this time did not substantially change from that of the Greeks. Owing to a lack of effective treatments, a physician's function was still primarily that of care and comfort.

The juncture that marks the true turning point in the evolution of medical thought is, of course, the discovery of germ theory—the understanding that microorganisms can cause disease. Lister, Koch, Pasteur, and others fundamentally transformed medicine from an art to a science. By the late nineteenth and early twentieth centuries, accurate diagnoses and successful cures could be applied to a wide range of specific diseases. From these developments, our contemporary "magic bullet" notion emerged, the "well nigh universal belief" that for every disease, we would be "capable of reaching and destroying the responsible demon within the body of the patient."[12] Optimism over the potential accomplishments of medicine flourished.

Medical advances accumulated after the discovery of germ theory. New methods of antisepsis made possible the prevention of infection and the likelihood of successful surgery. Complex chemotherapies evolved, and immunization became the "miracle" of modern medicine. Eventually, germ theory made medicine one of the most successful applied technologies in history. Long-standing disease were minimized; maternal and infant mortality rates dropped dramatically; life expectancy continued to increase.[13] Medicine replaced religion as the most respected and sanctified discipline of the modern world.

These successes, however, have also been the very causes of the alienation now seen as widespread in modern medicine. In making the disease process explicable, germ theory irrevocably altered the medical vision of the human entity. Germ theory signified, in the words of Andrew C. Twaddle and Richard M. Hessler, "the rise of germs and the fall of people."[14]

With its new scientific perspective, germ theory medicine now began to approach the patient less as a human being and more as an object of study. The holistic emphasis of the Greeks disappeared, and specialization increased. Doctors increasingly focused on the microorganism or the diseased organ and less on the comfort of the person. Over time, this redirection in medical philosophy has led to serious concerns about the rise in medically induced alienation. In direct response to this problem, new developments in medicine have evolved, what some have called the "post–germ theory" agenda.

NEW TRENDS IN MEDICINE: THE POST–GERM THEORY AGENDA

A variety of approaches have emerged in recent years which seek to deliberately direct medical attention back to the human and away from the technical, mechanical, and interventive. These movements may be considered countercyclical; in an era when medicine has become so fully scientized, these efforts seek to reassert that which is natural over and above that which is artificial, that which is communicative over that which is technological. In this way, it is thought, some of the alienating aspects of medicine may be counteracted. Some of the most significant of these post–germ theory strategies have revolved around the issues of birth and death, the two ends of the life cycle. Other developments address the fundamental focus of medical research and problems of social interaction.

In the area of childbearing, "natural childbirth" has seen a major resurgence both in practice and as ideology. The reemergence of this less medicalized approach has been aimed directly at what is considered the contemporary estrangement of the birth process. Richard W. Wertz and Dorothy C. Wertz, echoing Marx, write of this contemporary experience: "Hospital delivery had become for many a time of alienation—from the body, from family and friends, from the community, and even from life itself."[15] Similarly, Roger Shinn asks: "Are there points at which our contrived manipulation of ourselves and each other dehumanizes us? To some extent the attraction of 'natural childbirth,' for example, is the recapturing of an inherently human experience that got lost in the apparatus of hospitals and the dulling of consciousness by drugs."[16]

The demand for "death with dignity" has also been a direct response to the problem of estrangement in the dying process. To Elizabeth Kubler Ross and other leaders in this social movement, our present needs are not for greater technology, but for a more humane approach to the treatment of death. To implement this process, "living wills" and hospices have been instituted to give individuals greater control over their lives. Death, it is argued, is a *human* event and should be treated as such.[17]

Another post–germ theory development, in much the same direction, is the desire to shift medical attention away from a concentration on acute diseases and toward those that are more chronic or disabling, thereby minimizing the focus on intervention.[18] In particular, this would mean a new medical focus on health problems of the elderly as well as on those with long-term disabilities. This would entail greater attentiveness to prevention, rehabilitation, and public health concerns of a "low-tech" nature. The former concerns of Greek medicine again would be central, including the quality of our air, water, and environment, our life-style, nutrition, and exercise.

A final post–germ theory concern is the reassessment of medical interaction. Since the early 1960s, and probably coinciding with the disappearance of the "house call," there has been an enlarging critique of the physician's orientation toward the patient. Erving Goffman, for example, in 1961 argued that the prevailing medical model is one of "non-person treatment" in which "the patient is greeted with what passes as civility, and said farewell to in the same fashion with everything in between going on as if the patient weren't there as a social person at all, but only a possession someone has left behind."[19] These problems have continued to this day. Common charges against the medical establishment include the treatment of patients as if they were mere appendages to a machine, lack of human feeling, loss of bedside manner, overzealous intervention, withholding of clinical information, and lack of sensitivity and/or brutality in the discussion of diagnosis and prognosis, with both family members and patients.[20]

Efforts to effect real changes in medicine have been considerably fewer than the extensive discussions of the issue, but there have been some important breakthroughs. One is a new emphasis on "truth telling" and "informed consent," buttressed by changes in medical philosophy, medical curricula, and hospital ethics committees. Hospital ethics committees also have been instrumental in creating training programs for physicians on "ethical" dilemmas. Moral and social problems of patients' rights and autonomy have frequently come under discussion as well as sensitization to minority and ethnic relations and patient and family needs.[21]

It is too early to determine whether many of these new developments will become entrenched in the total context of our modern technological medical enterprise. Yet it would seem that these avenues do not represent mere fads. Natural childbirth effectively competes against local anesthetics for women in low-risk categories. Rejection of extraordinary measures and living wills for the treatment of the terminally ill are increasingly familiar in hospital settings, as is hospice care. Family practice is emerging as a major subspecialty, taking into account considerations of life-style and living conditions as diagnostic issues. The trend toward more holistic health care in the form of nutrition, exercise, and environment is widespread. Medical schools are slowly instituting social science and humanities courses into their curricula. Overall, efforts to counterbalance the alienating effects of medicine are gaining momentum.

The question that arises is whether this new momentum can be maintained. Through genetic and reproductive engineering, will new modes of alienative medicine emerge, such that the overall impact of more humanistic developments are undermined? In Chapter 3, it was argued that new modes of stigma are occurring at the very time when efforts are being made to ease their effects; in Chapter 4 new genetic interventions were examined for their effect on the powerless. Here, too, we must consider whether new dimensions of alienation are evolving at the very time we are seeking to counteract their negative social consequences.

Three areas of alienation will be examined: the human being as a *commodity,* that which can be bought and sold; the human being as *object,* that which can be manipulated; and the human being as *artificial,* that which is distanced from the natural. Each of these elements are significant dimensions in the context of modern alienation.

THE HUMAN BEING AS COMMODITY: MEDICINE AND THE MARKETPLACE

Richard Titmuss, in his book *The Gift Relationship: From Human Blood to Social Policy*, provides a useful starting point in our discussion of "commodity" relations.[22] Titmuss's investigated international blood collection systems of the 1960s, focusing primarily on England and Wales and the United States. Titmuss's comparative approach is instructive, for he found vast differences between the more commercialized blood system of the United States and the voluntary blood donor system of Britain.

In the United States in the 1960s and through the early 1970s, the blood collection system was primarily private and for-profit. In Britain, by contrast, the National Health Service ran the blood donor system as an entirely voluntary effort. To Titmuss, the overwhelming inferiority of the private, for-profit system was indisputable.

> From our study of the private market in blood in the United States we have concluded that the commercialization of blood and donor relationships represses the expression of altruism, erodes the sense of community, lowers scientific standards, limits both personal and professional freedoms, sanctions the making of profits in hospitals and clinical laboratories, legalizes hostility between doctor and patient, subjects critical areas of medicine to the laws of the marketplace, places immense social costs on those least able to bear them—the poor, the sick and the inept—increases the danger of unethical behavior in various sectors of medical science and practice and results in situations in which proportionately more and more blood is supplied by the poor, the unskilled, the unemployed, Negroes and other low income groups and categories of exploited human populations of high blood yielders.[23]

Owing to Titmuss's findings and other subsequent research, the United States, in the past decade, has shifted toward a significantly increased voluntary blood program and a greater limitation of the commercial market in whole blood. But the essential argument Titmuss advanced is still with us. When we commercialize medicine and its associated functions, we undermine the fundamental humanity of the system: we increase its alienative and exploitive components. As Titmuss argued, "If blood is considered in theory, in law, and is treated in practice as a trading commodity, then ultimately human hearts, kidneys, eyes, and other organs of the body may also come to be treated as commodities to be bought and sold in the marketplace."[24]

To Titmuss, the commercialization of blood erodes "altruism" and destroys the sense of community. Just as Durkheim argued that social tolerance is of a piece, so too did Titmuss argue that altruism is of a piece. "It is likely," Titmuss wrote, "that a decline in the spirit of altruism in one sphere of human activities will be accompanied by similar changes and attitudes, motives and relationships in other spheres . . . the consequences are likely to be *socially pervasive*. . . . If the bonds of community giving are broken the result is not a state of value neutralism. The vacuum is likely to be filled by hostility and social conflict."[25] To both Durkheim and Titmuss, there is an interdependence in social attitudes: push an important issue too far in a particular direction and other areas will inevitably follow.

Genetic and reproductive engineering techniques now make possible many new modes of commercialization. Surrogate parenting is perhaps the clearest example. Major concerns have arisen over women "selling" their bodies. As in the selling of blood, the pattern in surrogate parenting has been and likely will continue to be women of the lower classes selling their bodies to those of the upper classes. In Chapter 4 this problem was discussed in terms of issues of power and social stratification, but beyond these considerations are the subsequent alienative implications.

Surrogate parenting creates the opportunity to sell the human body, for the profit of the surrogate parent or for the profit of third-party brokers. Like prostitution and slavery, such practices increase the likelihood of denigrating the uniqueness of the individual. They may have equally debilitating effects on the community through the alteration of social norms. The sale of infants and children is currently forbidden in practice or in law in most societies. In surrogate parenting, however, payments are presumably for "services rendered." Yet the net effect is similar. Both women and children are ultimately treated as marketable commodities.[26]

Increasingly, women's groups are voicing serious opposition to surrogate parenting. In an "amicus brief" filed in the Baby M case in 1987, a group of leading feminist scholars, writers, and activists sought to overturn a Superior Court ruling that surrogacy contracts are legal. The brief argued that "as technology develops, the 'surrogate' becomes a kind of reproductive technology laboratory. . . . In short she has been dehumanized and has been reduced to a mere 'commodity' in the reproductive marketplace."[27]

Titmuss argued in *The Gift Relationship* that the selling of blood might ultimately set a precedent for the selling of other organs. So, too, the selling of one's body for reproductive purposes opens the possibility that other areas of genetic and reproductive engineering may develop into economic transactions, and indeed this already seems to be the case.

In 1985, in Victoria, Australia, for example, serious concerns arose over an in vitro fertilization proposal that included plans for a commercialized center. At Monash University, the site of the clinic, "fifty faculty members and others . . . attacked the proposal as leading to a 'new era of reproductive exploitation'" and "creating babies for profit-making purposes."[28] Indeed most reproductive technologies are now bought and sold on the marketplace, and the line between this practice and the marketing of infants for purchase is becoming increasingly hazy. Similarly, genetic engineering firms have become vast profit-making corporations. The creation of patents for life forms is perhaps the ultimate metaphor for life—human or animal—as a commodity.

The opportunities to shift the creation of life into a commodity forum are increasing with the development of genetic and reproductive engineering. In the process, we will perhaps inevitably venture into the territory Titmuss mapped. We will generate public policy that legitimizes the marketing of human beings or their parts.

In Titmuss's time, prior to the advent of most of these new technologies, he considered the sale of blood to be "one of the ultimate tests of where the 'social' begins and the 'economic' ends."[29] The solution to the dilemma Titmuss raised was to create a voluntary or nonprofit blood system. The problem was considered so vital that this course was in fact taken by the United States, at least in the collection of whole blood.[30] It is useful to consider the parallel. Where these new technologies are not used for profit, at least some of the more socially threatening considerations may be modified. To the degree that they become more firmly entrenched in the marketplace, however, their alienative properties are maximized.

THE HUMAN BEING AS OBJECT: NEW MODES OF MANIPULATION

Science deals with things not people.

—MARIE CURIE[31]

One of the most frequent and vociferous complaints against developments in genetic and reproductive engineering is the perception that human beings are increasingly *used* for purposes of scientific research. The new technologies appear to permit or encourage the objectification of the human being, as a thing or object for experimental purposes. Further, it is argued, if human beings are to be manufactured, made, or constructed, they may also be subject to all the types of manipulation that can be logically imposed on any manufactured item. Most of the new genetic and rcproductive technologies illustrate these problems.

The early experiences of in vitro fertilization demonstrate many of these fears. IVF scientists Robert Edwards and Patrick Steptoe conducted extensive prior research before the birth of Louise Joy Brown, yet they could not have known what would be the final outcome of their experiments. It has been noted that in their early work, several fetuses that resulted in miscarriages had serious physical disabilities. The scientists, however, did not notify the press or medical establishment for fear that their research might be halted. Instead they persevered, with the knowledge that what they were undertaking involved considerable risk to the woman and the fetus.[32]

The first child born of IVF, as it turns out, was healthy, as have most such infants born since that time. The point, however, is that the risk was willingly taken for scientific purposes. Philosopher Paul Ramsey was one observer who was deeply incensed at the experimental aspects of this research. Prior to the first success, he went so far as to ironically issue the "macabre hope" that the first infant born of in vitro fertilization might be disabled. In this way, he argued, the society would be forced to confront the potential peril of manipulating human beings in this manner.[33]

Patricia Spallone points out that "IVF is becoming more experimental with time, not less."[34] Powerful new drugs are being tried on women throughout the world, in an effort to fine-tune IVF procedures. One experimental drug, Buserelin, for example, "blockades" the normal functioning of a woman's pituitary gland to give IVF practitioners greater control over her bodily processes. The purpose of using the drug, at one site in England, was to test a hypothesis concerning follicle development in the ovary.[35]

The manipulation of human material is similarly displayed in the use of "spare embryos" which often emerge as a residue of in vitro fertilization techniques and other reproductive procedures. The use and propriety of handling spare embryos has spurred intensive debate internationally.[36] Research on spare embryos has been banned in some countries, for it is argued that unlike sperm and egg handling, the manipulation of embryos involves a potential human being. Again, Robert Edwards, the IVF specialist, has defended embryo research. He stated: "These spare embryos can be very useful. They can teach us things about early human life which will help that patient and other patients. It is very important to know that the types of patterns of growth that we get in our cultures are normal—as normal as we can ensure—and I believe it is absolutely essential to examine the speed of growth of these embryos . . . to examine the chromosomes . . . which involves flattening them on a tiny piece of glass and looking at the number of chromosomes they have."[37] While it may be true that spare embryos are "useful," the manipulative calculus in Edwards words is quite clear. To many, such scientific distancing from the human potential in those embryos is disquieting. It suggests that the willingness to deploy human material for experimental purposes is a prevailing and perhaps expanding medical and scientific paradigm.

The freezing of ovum is another example of manipulation for

scientific ends. The two scientists who claim to have originated the procedure publicly reported they did *not* perform an amniocentesis procedure on their first successful case for fear of inducing a miscarriage. They argued that they *wanted* the child to come to term, *even* if seriously disabled, so that they would be able to study the outcome.[38] This was obviously not an assertion of the rights of children with disabilities, as discussed in Chapter 3, but rather what appears to be the protection of research property. In the eyes of these scientists, this fetus seems to have been fundamentally an experimental entity.

The overriding concern is that the development of these techniques will inevitably require the extensive use of human beings as "guinea pigs." However, it also may be argued that *all* first applications of scientific procedures use human beings in this way. Were not the first recipients of the polio vaccine individuals who were being *used* for scientific purposes? The same could be said of every new development in medicine. Is genetic intervention different?

The essential distinction lies in the purpose of the research. For example, several years ago Amatai Etzioni drew a distinction between technologies that are recognizably therapeutic and those used for other purposes.[39] It seems likely, for example, that developments in genetic surgeries that seek to *cure* diseases will be perceived as less objectifying. Similarly postnatal PKU screening, in which treatment can be prescribed after identification of the disease, is also a fundamentally familiar therapeutic approach. Other procedures, however, seem to be occurring much for the sake of scientific curiosity. They will ultimately provide benefits for only a very few people, and as research, they will have to use as guinea pigs people who are not ill or even threatened by illness.

Most of the new reproductive technologies such as in vitro fertilization, cloning, artificial insemination, artificial wombs, embryo transfer, and surrogate parenting are examples of procedures on the non-ill. The prospective parents may have fertility problems, but they are not sick. If the classic premise of medicine is "Do no harm," intervention in these cases is at best on shaky grounds. Although the prospective parents may provide informed consent and are willing guinea pigs in the process, the pattern that is being established is one of manipulation for the purposes of scientific endeavor.

As these technologies multiply and their potential enlarges, the acceptability of this thought process may extend to other areas. Such interventions may fundamentally enhance the alienative dimensions of medicine. The increasingly common manipulation of spare embryos and the rising use of experimental means of reproduction suggests that a willingness to *use* human beings or potential human beings as objects for investigative use is gaining momentum.

THE HUMAN BEING AS ARTIFICIAL: THE LOSS OF NATURE

> *Art, artifice, artificial. . . . The world that is being created by the accumulation of technical means is an artificial world. . . . It destroys, eliminates, or subordinates the natural world, and does not allow this world to restore itself.*
>
> —JACQUES ELLUL,
> The Technological Society[40]

Contemporary developments in "natural childbirth" and "natural death" suggest an image of the natural world. They seem to assume that there are natural modes of behavior which we ought to abide by, but which we too often violate. If we violate these more natural behaviors, it is argued, we move away from that which is intrinsically human and healthy. If we obey them, our humanity is maintained, our physical and psychic life is enhanced.

This desire to ward off the artificial and to dignify the natural is a recurring theme in history, literature, and social thought. The romantic movement of the late eighteenth and early nineteenth centuries is but one example. In response to the industrial revolution, English poets such as Wordsworth, Byron, Keats, and Shelley expressed a desire to preserve nature; to reassert the superiority of the natural over human and technological intervention. That yearning is still with us.

Clearly, new technologies such as artificial insemination, in vitro fertilization, frozen embryos, artificial wombs, and surrogate parenting, to name but a few of the new approaches, seem to stand at precisely the opposite point from what we might consider natural. It is therefore not surprising that at the very time so many new artificial methodologies are being developed, there is also a surge for that which appears more natural.

To some scholars, many of the new technologies seem so alien from that which we consider natural that intense debates have emerged over the very meaning of the terms *artificial* and *natural*. One debate poses the question this way: Should reproduction be left to its original form, as "nature intended," or is laboratory reproduction in fact equally natural, and perhaps more efficient and healthier, giving human beings new options, choices, and hopes? Is it more human to create, build, and control nature, or instead, should human reproduction be one of the last "sacred" areas, which science and technology ought to leave alone?

This debate has its extreme sides represented by Joseph Fletcher and Leon Kass. Joseph Fletcher, writing for the interventive or artificial approach, writes that the domination of nature is distinctively human; it is in fact the very trait that distinguishes the human species. "Man is a maker and a selector and a designer," according to Fletcher, "and the more rationally conceived anything is, the more human it is. Any attempt to set up an antinomy between natural and biological reproduction on the one hand and artificial or designed reproduction on the other is absurd. . . . It seems to me that laboratory reproduction is radically human compared to conception by ordinary heterosexual intercourse. It is willed, chosen, purposed, and controlled, and surely these are among the traits that distinguish Homo Sapiens from others in the animal genus, from the primates down."[41]

The alternative perspective advances nature as a *wiser* choice. The evolution of human beings, this side argues, has given humankind proven methods for preserving the species and providing love and care to children and family. Leon Kass, for example, one of the strongest proponents of this perspective writes: "Is there perhaps some wisdom in the mystery of nature, which joins the pleasure of sex, the communication of love, and the desire for children in the very activity by which we continue the chain of human existence? Is not biological parenthood a built-in 'mechanism' selected because it fosters and supports in parents an adequate concern for and commitment to their children? Would not the laboratory production of human beings no longer be *human* procreation?"[42] Kass poses the possibility that the biological revolution, over time, will violate "even the *distinction* between the natural and the artificial."[43]

Both arguments are troublesome. Although Kass has a vision of what he considers natural, we need to question whether any truly natural modes of behavior actually exist. What is natural to Kass may mean no more than that behavior to which we are most accustomed or that which we most idealize. Both childbirth and death may be viewed as socially constructed.[44] As Ann Oakley writes, "Pregnancy and birth, like death, are bodily experiences which have 'negotiated realities' in different cultures."[45] Ruth Hubbard is even more emphatic:

> I do not believe that any human pregnancy or birth is simply "natural." All of us live in societies that define, order, circumscribe, and interpret our activities and experiences. Just as our sexual practices from earliest childhood are socially constructed and not a natural unfolding of inborn instincts, so are the ways of structuring and experiencing pregnancy and birth. Whether a woman goes off to give birth by herself (as do Kung women in the Kalahari desert), calls in the neighboring women and perhaps a midwife (as my Grandmother did in a small town in eastern Europe), goes to a lying-in hospital (as I did in Boston some twenty-five years ago), or has a lay midwife come to assist her and her partner at home (as several of my friends have done in the last few years), all these are socially devised ways, sanctioned by one's community, even if sometimes not by the medical profession or the state.[46]

Fletcher's Pollyanna visions of reproductive technologies and laboratory control are similarly unresponsive. To simply herald the coming of technologies, christen them "progress," and not acknowledge their deeply alienating potential fails to address contemporary medical dilemmas. The control of nature may take many forms, some of which may be humanizing or empowering, as he notes, but some of which also may be deeply alienating. Ultimately, it may be less the degree of artificiality that is at stake, than the *imposition* of technologies that is most threatening and most alienating to patients. It is perhaps not control of nature that is at issue, but control by patients.

Whether there are aspects of birth and death that are truly natural is clearly open for debate. Whether the control of nature gives us power is equally open for debate. But what seems undebatable is the growing distrust and anguish in our experiences with medical events. The controversy over nature versus artifice, though arguably muddled in its terminology, does nevertheless point to a genuinely felt concern on the part of patients. The "natural childbirth" movement emerged as a rebellion against the loss of control and the experience of social distance in the medical process. The same may be said of the "death with dignity" movement. Both movements, as well as the continuous yearning for greater simplicity and for home environments rather than sterile hospitals, are indicants of widely felt grievances against the medical system and the administration of medical technologies. If birth and death are social constructions, we are constructing them poorly. Rather than creating ways to comfort, we frighten and often terrify patients. We make critical life moments—birth and death—periods of fear and loneliness, surrounded by strangers and machines. The most vital concern that the nature-artifice dilemma points to is the need for change and reform. Yet if new genetic and reproductive technologies continue to emerge, demanding more not less of the very environments that alienate, those needed changes will be seriously impeded.

CONCLUSIONS

The social history of medicine reflects an increasing concern with the problem of alienation. Over time, medicine has come to be perceived as distancing the patient from the processes of diagnosis and treatment. New policies in medicine have sought to counteract these alienative properties, but like many such developments, they represent tentative and fragile first steps along a path that is overwhelmed by opposing forces.

Developments in genetic and reproductive engineering represent new modes of generating medical alienation. Their propensity to *commercialize* the human being, *objectify* the physical body for purposes of experimentation, and introduce *artificial* methods of creation suggests that the alienative directions of medicine will likely predominate and multiply in the coming years.

The hard-won gains in contemporary medicine, which have sought to humanize the medical enterprise, recede into insignificance in comparison with the new technological breakthroughs of genetic and reproductive engineering. To maintain the "patient as person" at the center of this medical worldview will be an increasingly difficult task in the years ahead.

If individuals are to benefit from medical care, they must not fear its social institution. As currently constituted, the medical establishment seems overwhelmingly anonymous and impersonal. It is stratified by class, in regard to choices and quality, and distinguished by an orientation that tends to rank technological output more important than human interaction. The gains achieved through informed consent, truth telling, hospital ethics committees, and a sensitization to individual needs are small, but if further developed, these approaches might become significant parts of a larger institutional reorientation.

Jacques Ellul has written that a "nation that has reached a pitch of perfection in its technical organization sometimes feels this perfection to be intolerable."[47] Inevitably we will have to consider balancing the myriad of choices made available by science, technology, and medicine with a careful evaluation of the social context in which these choices are offered. Returning medicine to a human scale is not an impossible task. It may take additional resources, certainly additional time, yet the necessity is unquestionably there.

Part III

Tempering Genetic Welfare

Thus far we have explored the impact of Genetic Welfare on three social problem areas: stigma, powerlessness, and alienation. We have argued that a genetic worldview creates new social costs for those who are vulnerable. New stigmas are emerging, new inequalities are produced, and new modes of social distance are generated. Because of these problems, it is important to consider how we, as a society, have sought to temper or moderate the impact of genetic and reproductive technologies. What controls have been sought to prevent the negative or potentially negative consequences of new genetic and reproductive technologies?

To address this problem of control, Part III of this book will examine issues of public policy. We will begin by assessing regulatory policies created by scientists or government or a combination of the two. We will next look at regulatory situations emerging from the "bottom up," from citizen movements and citizen responses. Finally, we will examine the public agenda itself and its relative responsiveness to genetic and social concerns.

The fundamental dilemma that will be addressed in each of the following chapters is one of balance. Diverse interests are at stake in the formation of public policy. For example, freedom of inquiry is usually weighed against protection of the public. Citizen participation is typically balanced against professional privilege. Genetic strategies may be considered in light of their costs to a social agenda. There is little consensus and much controversy in our society over the relative importance of these different concerns. Opinion varies greatly, depending on one's location in society, one's professional responsibilities, and one's knowledge of the problems involved.

Because of the disparity of opinion on these issues, it is important that we begin to examine the structures that are now in place and the regulatory policies that have thus far emerged. We need to consider how well we have done and what still needs to be accomplished. Our success in managing genetic technologies will determine the costs and benefits that will accrue to different groups. How well our social policy is formulated will determine the impact of Genetic Welfare on those who are most subject to its problems.

6

Science, Regulation, and Public Policy

No man is allowed to put his mother into the stove because he desires to know how long an adult woman will survive at a temperature of 500 degrees Fahrenheit, no matter how important or interesting that particular addition to the store of human knowledge may be.

—GEORGE BERNARD SHAW,
The Doctor's Dilemma

Haven't we already faced the blunder in a rather remarkable way, not with genetic manipulation, but in treatment, with something like thalidomide, and survived the shock? In other words, a mistake was made, despite all the pre-testing and planning. It was an unprecedented disaster in medical science, but we don't scrap the whole program. We all respond to it in different ways—embarrassment, chagrin, despair to a certain extent. The first blunder made in genetic manipulation cannot be any more or less of a disaster than the kinds that have already been made. In other words, we can live with the problem, and we must live with the problem of making mistakes. That is part of the trial and error system. That is much of what science is about.

—NEIL TODD,
Symposium on the Identity and Dignity of Man, sponsored by Boston University and the American Association for the Advancement of Science

The willingness to accept sacrifices for the sake of potential scientific discoveries has long been a trademark of those in the sciences. Perhaps only a few skeptics, such as George Bernard Shaw, may question the enterprise entirely. But scientists often view the process in cost-benefit terms: to make gains, science must experience setbacks. Individual subjects may be lost in the process; their suffering is a regrettable but perhaps inevitable outcome of the process of experimentation. This is, as Neil Todd argued, "part of the trial and error system. This is much of what science is about," and to many scientists, its wonders are worth the costs.

The benefits of science and the intrigue of exploration have been commemorated far more often than the losses, particularly by those who have excelled. Einstein spoke of the "eternal riddle" whose contemplation "beckoned like a liberation"; Newton wrote of probing "the great ocean of truth"; Watson and Crick turned science into an adventure saga.[1] The challenge of solving great scientific and technical problems is part of the mystique of our modern world.

The allure of the scientific enterprise and our intense belief in its contributions to contemporary life have granted the pursuit of science many privileges. Among these, in most Western societies, is "freedom of inquiry." Nowhere is this an absolute right, but it is one that tends to be limited cautiously, and only when there is much confirmation of potential harm.

In this chapter we will consider the historic tensions between freedom of inquiry and protection of the public. We will look at the idea that some knowledge may be harmful in and of itself and examine the concept of regulation. More specifically, we will assess a variety of relatively new mechanisms that seek to manage decision-making processes in medicine and science, including such structures as biosafety committees, medical ethics committees, institutional review boards, and presidential commissions.

Most important, in this chapter we will consider whether the thrust of the new regulatory mechanisms may be used to balance the competing claims of Genetic Welfare versus Social Welfare. Can the encroaching problems of stigmatization, powerlessness, and alienation be tempered, at least to some degree, by public policy? Are the structures that we now have adequate to the task of oversight; do they incorporate considerations of social as well as physical harm? These questions will be critical components on our policy agenda in the coming decades.

LIMITING SCIENCE: THE DILEMMA

Historically, in the United States, we have approached the limitation of science gingerly, both in thought and in action. Regulations, when they existed, were fundamentally suspect and regarded as a necessary evil. When created, they were virtually always cases of last resort. More recently, this supportive attitude has changed. A new wariness has emerged, as the press and various regulatory agencies have begun to reveal cases where scientists stepped "over the line." The most important of these revelations was the documentation of the Tuskegee syphilis study.

The Tuskegee syphilis study began in the 1930s. Scientists and physicians of the U.S. Public Health Service of Macon County, Alabama, sought to explore the long-term effects of untreated syphilis. Using primarily poor black sharecroppers as subjects, the experimenters transformed them into human guinea pigs. They never informed the subjects of the nature of their disease and at various points actively worked to prevent their treatment. They permitted the subjects to suffer the continued effects of syphilis long after penicillin was known to be an effective medication.[2]

This study was continued for forty years until its existence was revealed in the national press. It was not until 1972 that an investigation was conducted by the Senate Subcommittee on Health. As a minimal compensation for their suffering, reparations were granted by the United States government to the few surviving members of the study and to the families of those who were deceased.[3]

The exposure of the Tuskegee study caused major reverberations throughout the United States; its parallels to the Nazi medical experiments were cited by many.[4] It was noted that science may easily violate basic tenets of humanity, irrespective of culture or politics. The Tuskegee syphilis study now stands as a symbol of contemporary scientific abuse.

Other cases of negligence have contributed to a new sense of caution about the process of scientific research. One was the Willowbrook study of the 1950s in which physicians injected institutionalized and disabled children with live hepatitis virus.[5] Similarly, the congressional Office of Technology Assessment points out, "The U.S. mass media had in the 1960s carried a number of reports about unsavory situations—here and abroad—in which prisoners, children, the poor, and the elderly were exposed to unwarranted risks in the name of 'experimentation'."[6] More recently, the space shuttle *Challenger* disaster, in 1986, also illustrated to the public the willingness of scientists to risk individual lives in the face of uncertain knowledge.

The outcome of all these events has been an effort to impose at least some boundaries on the exercise of scientific research. A variety of guidelines for the protection of human subjects are now emerging. Moreover, not only are the methods of science under scrutiny, so are the very areas of investigation. In the past few years a variety of efforts have been undertaken to consider the paths scientists have chosen and the effects new scientific knowledge may have.

Some critics, for example, have gone so far as to suggest the idea of "forbidden knowledge," or what Robert Sinsheimer calls "inopportune knowledge."[7] Sinsheimer poses the possibility of giving science a "disciplined direction," because, he says, certain forms of knowledge, "at a given time and stage of social development," may be "inimical to human welfare—and even fatal to the further accumulation of knowledge."[8]

Sinsheimer warns that once knowledge becomes part of our understanding, its impact is irreversible. To illustrate, he suggests an elaboration of the "uncertainty principle": since we believe it to be true that the observer inevitably changes that which is observed, it also may be true that what is observed may irrevocably change the observer. This and other variations on Pandora's Box—that the spirits, once released, can never be returned—have become significant points of departure in much recent debate. They have led to a consideration of regulatory measures, efforts to contain certain kinds of knowledge or action.

REGULATION: WHAT IT IS AND WHAT IT ISN'T

Regulation is a thorny issue in the United States. To some it raises the specter of red tape, government interference, or the limitation of freedom. For others, it is a needed, although arduous path to public protection and reform.

Examples of regulatory bodies include the Food and Drug Administration, the Occupational Safety and Health Administration (OSHA), the Anti-Trust Division of the Department of Justice, the Environmental Protection Agency, and the Office of Civil Rights. All these agencies have their supporters and detractors.

One of the nagging difficulties in assessing the effectiveness of regulatory bodies is that in the pursuit of their missions they often seem to please no one. While some argue they are too soft, others complain of their intrusiveness. Of the Office of Civil Rights, for example, Jeremy Rabkin writes:

> It is hard to find anyone who will defend the record of HEW's Office for Civil Rights (OCR) over the past decade; it is equally hard to find any agreement on the nature of its faults. Many critics see OCR as a hotbed of regulatory zealots obsessed with vast social engineering schemes that bear little relation to their actual statutory mandates. To its constituents, on the other hand OCR is a lumbering bureaucracy, addressing its obligations in the most timid, half-hearted and ineffectual manner.[9]

The fundamental purpose of regulatory bodies is to protect interests that may be typically overlooked, for example, public safety, the environment, the role of minorities, and so on. In an ideal setting, businesses, corporations, and scientific groups would automatically take these interests into account. Since this is obviously not the case, we have created ways to impose our concerns. The outcome, however, is often slow, cumbersome, and piecemeal, but it appears to be the only approach we now have available to enforce these claims.

Regulation seems to work best when the interests of those who might be harmed are fully represented. When those who have grievances or potential grievances are part of the process, the approach seems just. Alternatively, self-regulation, where there is no representation of other interests, is troubling. The absence of other parties calls into question the range of considerations that are raised and the nature of the vested interests that may dominate decision making.

Regulations for genetic and reproductive technologies have emerged in a variety of different forms, but interestingly, the most famous case of regulation, that of recombinant DNA, began and ended with scientists regulating themselves. In the next section, we will consider this case of regulation and its implications for public policy.

A CASE STUDY IN SELF-REGULATION: THE RECOMBINANT DNA DEBATE

The recombinant DNA debate began in June of 1973 when scientists at the Gordon Conference on Nucleic Acids opened discussions of this newly emerging technology. Scientists at this conference raised the possibility that through research into recombinant DNA, a "biohazard," that is, an altered genetic organism, might be released into the environment causing serious harm to human, plant, or animal life. Of particular concern at this time was research done in cultured *E. coli,* a bacterium normally present in the human intestine.[10]

Following the Gordon Conference, two scientists, Maxine

Singer and Dieter Soll, wrote a letter to the National Academy of Sciences and the Institute of Medicine, calling for a study to investigate the possible risks of recombinant DNA. The letter was published in the journal *Science,* in order to alert the scientific community to the perceived problem.[11]

The National Academy of Sciences then asked Paul Berg of Stanford University to head an investigatory committee. The result was a letter, once again published in *Science,* that called attention to our lack of knowledge regarding recombinant DNA biohazards. The Berg Letter, as it came to be known, requested that scientists halt their investigations of recombinant DNA until the dangers were better understood. The letter also recommended that an international assembly of scientists meet to consider potential risks.[12]

In 1975, soon after the publication of the Berg Letter, scientists involved in research on recombinant DNA met at a specially convened conference in Asilomar, California to discuss the idea of biohazards. At this historic conference scientists took an unprecedented step by committing themselves to self-regulation. They created a set of safety guidelines and, most important, declared a moratorium on certain forms of genetic research until more was known about the risks. Following upon the lead of the scientists, the National Institutes of Health (NIH) also entered the picture, issuing its own set of guidelines paralleling those of the scientists.[13]

As it turned out, the moratorium was brief. As more information became known on recombinant DNA and *E. coli,* scientists came to believe that their initial concerns were overstated. They soon relaxed their controls over recombinant DNA. NIH then similarly revised and simplified its guidelines so that "no class of experiments" was "prohibited" and NIH approval was needed only for work involving the most serious potential risks.[14]

Thus this movement in self-regulation quickly came and went. The policy question this raises, illustratively, is whether such self-regulation is desirable. That scientists acted quickly, when concerned about a threat to public safety, is an important consideration. However, what is also noteworthy is how rapidly scientists dispensed with controls once they determined the immediate danger was past. Moreover, in acting unilaterally, the scientists effectively held off the action of others. Clifford Grobstein, for example, argues that Asilomar was a "master stroke. . . . Several scientists recognized a problem, took an initiative directed to the public interest, devised an interim solution, and succeeded in defending it against significant opposition that would have imposed far greater restriction or, in the words of the most militant opposition, 'shut it down.'"[15]

Public policymakers also were troubled by the relatively autonomous actions of the scientists. Senator Edward Kennedy, for example, then chairperson of the Senate Subcommittee on Health, argued: "It was commendable that scientists attempted to think through the social consequences of their work. It was commendable, but it was inadequate. It was inadequate because scientists alone decided to impose the moratorium and scientists alone decided to lift it. Yet, the factors under consideration extend far beyond their technical competence. In fact, they were making public policy. And they were making it in private."[16]

Senator Kennedy's argument raises several important questions. Who should control decision making? Should it be only physicians and scientists, or only elected officials, or should the public be involved, or some combination of them all? Of similar import are questions related to the goals of public policy. Should policy respond to problems as they develop, or function in a preventive and anticipatory manner? These are vital questions which we, as a society, have not yet begun to address.

It is likely that in the next several decades we will need to radically alter our perspective on regulation. Much more will be called for than we now have. The potential dangers, both to our physical being as well as to the fabric of society, will challenge us to create new mechanisms for regulating advances and moderating their effects. The models that we have created thus far are only tentative steps in this direction. These structures, as we will see, are somewhat haphazard, arising out of immediate needs and pressing concerns. They offer little in the way of planning and do not yet have the scope that might mediate the social as well as technical aspects of Genetic Welfare. Nevertheless, most are small steps in a positive direction.

The models we will examine include biosafety committees, medical ethics committees, institutional review boards, and presidential commissions. All are multidisciplinary, and all include public participation to a greater or lesser degree.

REGULATORY ACTION: BIOSAFETY COMMITTEES

Institutional biosafety committees emerged, as regulatory mechanisms, as a direct result of the recombinant DNA controversy discussed in the preceding section. They were created by NIH Guidelines in 1978 to oversee the safety procedures of recombinant DNA investigations.

These committees were required to maintain a combination of professional and "nonaffiliated" persons, which in some cases meant local community residents. In a 1982 study that examined a sample of these committees, Diana B. Dutton and John L. Hochheimer found that the requirement for public participation was "generally constructive."[17] More problematic, however, were the actual regulatory functions of the committees. Dutton and Hochheimer found that 96 percent of all research proposals were approved, 73 percent without any modification at all. Members of the committees sometimes saw themselves as "rubber stamps."[18]

One of the most interesting and important aspects of these committees is their declining significance over time. Between 1976 and 1978, primary responsibility for enforcement of guidelines was with NIH. After 1978, this responsibility was delegated to institutional biosafety committees. But, by the early 1980s, these committees were barely functioning; responsibility for decision making had returned almost entirely to the primary investigator, precisely where it had been prior to the controversy and guidelines. In fact, in 1981 a proposal to completely dismantle the committees was submitted to the NIH and narrowly defeated only by the protests of several congressman.[19]

These committees were formed for a very specific purpose. They emerged out of a particular fear of a biohazard enunciated by scientists themselves. According to the same scientists, these biohazards then turned out to be fundamentally unsubstantiated. Hence the role and purpose of these committees appeared to be superfluous, and the committees quickly withered in significance. Control returned to the scientists alone. Thus this rare occurrence of a decision-making structure, generated by scientists themselves, was a short-term phenomenon, reluctantly adopted and quickly shed when the immediate danger was past.

As a model for regulation or decision making, both in origination and over time, biosafety committees appear to have minimal viability. They have addressed themselves, at least to this point, only to physical dangers. It is interesting to note that, thus far, there has been little perceived need to oversee the broader or social implications of recombinant DNA technology. As a mechanism for fending off the social problems of Genetic Welfare, these committees appear to have only limited potential.

REGULATORY ACTION: MEDICAL ETHICS COMMITTEES

Medical ethics committees, sometimes called institutional or hospital ethics committees, were one of the earliest forms of multidisciplinary decision making to emerge in medicine and the sciences. They were begun in the 1960s as a means for hospitals to address abortion issues, prior to *Roe v. Wade*. By order of many State statutes at that time, a hospital ethics committee could determine if a woman's life was in danger and decide whether to permit an abortion. Soon after, these committees were expanded to address other considerations as well.

The most important impetus for the widespread emergence of medical ethics committees was the case of Karen Ann Quinlan in 1976. "In that landmark decision, the court quoted an article by Dr. Karen Teel which suggested that the way to improve medical decision making was for each hospital to establish an 'Ethics Committee composed of physicians, social workers, attorneys, and theologians.'"[20] Because of this unique ruling, the medical ethics committee in the Quinlan case was given a critical role in determining the right of the family to terminate treatment.

The role of medical ethics committees was further enhanced by their participation in the controversial Infant Doe case in Indiana in 1982 and the Baby Jane Doe case in New York State in 1983. In 1983, on the basis of these and similar cases, the Department of Health and Human Services issued guidelines to encourage but not mandate the creation of infant care review committees, a subcategory of medical ethics committees.[21] Many hospitals now have these committees, and they have become increasingly central in determining issues of treatment, nontreatment, and extraordinary care for newborns.

Perhaps the most controversial role for medical ethics committees occurred during the early 1970s, over the issue of kidney treatments. Before renal dialysis was fully subsidized by the federal government, medical ethics committees were required to select patients who would receive access to what was then only a limited number of dialysis machines. The process of choosing prospective patients included evaluating such issues as social standing, relative prognosis, age, potential contributions to the society, and other combinations of social and physical attributes. Because resources were so few, and the treatment so lifesaving, the committees were dubbed "God committees." Eventually, the life-and-death roles of these committees became a matter of public concern. So shocking was the nature of the selection process, to the public and Congress, that the concept of subsidizing renal dialysis came to the fore. In 1972, PL 92–603 was passed by Congress, to underwrite the costs of renal dialysis and make treatment more broadly available. Kidney disease thus became the first disease for which costs are almost fully subsidized. Medical ethics committees were then relieved of the responsibilities of patient selection.[22]

A more recent issue for medical ethics committees is the determination of recipients for organ transplantation. The committees' duties once again include patient selection, during which the committees sometimes function like the "God committees" of old. As with dialysis, national policy is once again under consideration; if implemented, it would alleviate the necessity of making such decisions.[23] In addition, many other issues are now subjects for decision making by medical ethics committees, including No Code or DNR (Do Not Resuscitate) guidelines, the determination of brain death, issues of competency, patients' rights codes, care of the medically indigent, and discharges under the DRG (Diagnostic Related Group) system.[24]

Unlike biosafety committees, the number of medical ethics committees is increasing rapidly. The rising prominence of bioethical problems in medicine appears to be making their presence more and more necessary, even in small community hospitals. Their structures, however, vary widely, and they often have differing degrees of decision-making authority. They are typically composed of a wide range of hospital personnel including administrators, physicians, nurses, and social workers. Others who are sometimes included are theologians, lawyers, and community spokespersons. Consumers of health care or lay persons with no special affiliation may also serve on the committees.

The benefits of these committees clearly lie in their focus on ethical concerns. They encourage and set aside a period of time during which a thoughtful assessment of medical dilemmas may occur. Unlike the normally pressured decision making of the day-to-day routine, ethics committees flag certain issues for special attention. Over time, if rational decision-making procedures are developed, they may contribute to the formation of hospital policy.

Of particular significance is the multidisciplinary nature of these committees. Because of the range of occupations involved, they have the potential to raise a wider range of issues than would normally occur in typical case-by-case deliberations. They may also contribute to the education of hospital personnel by engaging speakers, conducting symposia on controversial issues, and circulating information that is not routinely available.

The great difficulty with medical ethics committees is that standardization is missing. Each hospital has its own committee; each makes its own rules; and each must sometimes reinvent the wheel. When committees fail to work smoothly, decisions may be arbitrary and representation on committees may not be broad-based. Moreover, the lines of authority are often unclear, and the committees' roles vis-à-vis the courts have been inconsistent.[25] Although these committees may be extremely useful in resolving immediate and unique dilemmas faced by hospitals, what is often missing is a national perspective; this is to some degree intentional. As with biosafety committees, physicians and hospitals often see these structures as a way to keep the government and external regulations at bay.[26]

As a method for bringing significant medical issues to the fore, medical ethics committees represent a mixture of success and failure. On the negative side, they lack consistency and breadth. The arbitrary manner in which these committees are sometimes created and administered is a matter of national concern. On the positive side, medical ethics committees represent a unique effort to blend a variety of health perspectives. Most important, they constitute a first step in addressing the problems of technological change in health care. The social aspects of health care are often taken into account in their deliberations, and they have helped define newly emerging problems in medicine. They thus have the potential to deal with many aspects of Genetic Welfare. If they continue to develop in this direction, ultimately, they may serve as a model for more elaborated regulatory structures of the future.

REGULATORY ACTION: INSTITUTIONAL REVIEW BOARDS

Institutional review boards (IRBs) are designed to protect the human subjects of scientific research. Boards review proposals for research in light of their potential risks. They were first mandated by the Department of Health, Education and Welfare in 1974 and continue to be directed by the Department of Health and Human Services.

IRBs are required at all research institutions that receive federal funds, typically colleges, universities, teaching hospitals, research institutes, and research centers. They are staffed by members of each institution and thus have an institutional basis in the context of federal guidelines. As John A. Robertson writes, "This system is a unique blend of public incentive and private initiative, of public oversight and peer control, for it depends on research institutions to apply federal standards to the activities of their scientists."[27]

IRBs received their most important impetus from the exposure of the Tuskegee study cited earlier. Until that time regulations had been administered by the Public Health Service and NIH. In the light of the Tuskegee revelations, these regulations were found to be inadequate. The Department of Health, Education and Welfare then created its institutional review board policy which was later formalized by the National Research Act, passed by Congress in 1974.

IRBs stress several facets of safety and protection for human subjects. Among these, one of the most important is informed consent. Subjects must be fully informed of the risks entailed in their participation in research. Those not fully competent to provide consent, or whose choices are limited (e.g., prisoners, children, the institutionalized, and the poor), are groups of particular concern. Other important considerations include recruitment and compensation of subjects and coherence of research design.

Many criticisms of IRBs have arisen. These include an overemphasis on bureaucratic regulations, an inability to properly monitor projects, and occasional charges of political influence.[28] IRBs are sometimes difficult structures in which to work. They create multiple deadlines for researchers and impose somewhat arbitrary demands.

Nevertheless, the benefits of IRBs outweigh their costs. Like medical ethics committees, IRBs serve an important role by their very existence. They function as a reminder of ethical and protective concerns. The very fact that a research proposal must be approved by an IRB increases attentiveness to subject safety and dignity. It forces an enumeration of these concerns in the initial project proposal; ongoing monitoring may keep ethical issues in view.

IRBs have one advantage over medical ethics committees—they are overseen by standardized federal policy, even though administered locally. This seems a reasonable compromise. Moreover, there is some federal clout, albeit rarely used: federal monies can be withdrawn.

The IRB structure, though imperfect, functions as a basic anchor for ethical concerns in scientific research. Their most important contribution is to serve as a constant reminder that the interests of individuals and groups may not be sacrificed in the pursuit of scientific advance. Their concern with groups who are particularly vulnerable, such as the poor and minorities, focuses attention on those most subject to harm. Because of these concerns, IRBs could play a significant role in the future, in warding off the most negative effects of Genetic Welfare.

RESEARCH AND OVERSIGHT: PRESIDENTIAL COMMISSIONS

Throughout American history, presidents have appointed commissions to investigate and report on particularly thorny issues. These commission reports are typically advisory, and the president has the choice of accepting or rejecting their recommendations.

One of the most important commissions in recent decades was the President's Commission for the Study of Ethical Problems in Medicine and Biomedical and Behavioral Research. Appointed by President Carter in 1979, the commission completed its work in 1983 under President Reagan, issuing sixteen volumes covering ten major topics. These topics included dilemmas of health care such as access, informed consent, the definition of death, and treatment decision making. Also included were biomedical issues such as genetic screening and genetic engineering. Finally, the commission included several volumes on human research and the protection and compensation of human subjects.[29] These reports received much attention and laid the groundwork for many important debates in the biomedical arena.

The reports responded to many critical issues on the national agenda. A variety of groups were involved, and witnesses were drawn from diverse disciplines and occupations. They attempted to respond to issues troubling a broad range of public interests. Thus, for example, in one case a topic was taken up specifically at the request of the general secretaries of the Council of Churches, the Synagogue Council, and the United States Catholic Conference.[30] Their joint theological letter, addressed to the president, spurred an inquiry on genetic engineering of human beings.

All told, the reports amassed a wealth of data on critical ethical issues in the life sciences. The attention to detail and the range of witnesses and consultants attest to the care given to their deliberations. Further, there were a broad range of recommendations for each subject area, indicating important directions for change, growth, and in some cases limitation.

There is little question that these volumes represent a significant achievement. They received much attention in the media and served to inform the public and the academic community of important debates. The difficulty lies in their implementation. Who follows through in the decision-making process? Which recommendations will be enforced, by whom, to what degree, and with what funding? There is often no order of priorities and no overall plan.

This president's commission had some successes in implementation. Its recommendation for a uniform definition of death, for example, was quickly adopted by many state legislatures. This issue, however, had a long history of consensus preceding it.[31] The commission also provided a substantial impetus to medical ethics committees, particularly in situations where a patient is deemed incompetent.[32] Most of the other recommendations, however, remained essentially advisory.

A commission report, ideally, would be an interim stage, structurally designed to feed into an implementation agency. Without such a design, such reports often sit in the government documents sections of libraries, gathering dust. They provide significant data on which to base regulations, but they do not represent action components themselves.

Nevertheless, these commissions are important because they thoughtfully address issues in-depth, over a period of time. They serve as a roundtable for a range of interests that would not normally confront one another. As a first step in generating public policy, these committees play an important role. Like the Kerner Commission Report on violence in the cities in the 1960s, they summarize where we are at a given moment.[33] But much more is needed. It is not sufficient to point to problems. We must also formulate policy. This is difficult, because there is so little consensus among scientists and the public on how to proceed. In the next section, we will consider the response of scientists to the current range of regulations now in effect.

REGULATORY MECHANISMS: ACCEPTANCE AND RESISTANCE

> *I'm expressing civil disobedience. We can sit and talk about Dutch Elm Disease or we can do something about it. . . . I did what I did to save a year and perhaps to draw attention to the fact that we have to be a little more careful in terms of the restrictions we impose.*
>
> —GARY STROBEL, plant biologist, after releasing into the environment an altered genetic organism, without regulatory permission, 1987[34]

How well have the various regulatory models and decision-making structures been accepted by scientists and physicians? The data suggest that, overall, they have been accepted, but with considerable reluctance and impatience. In a few instances, such as the case of Gary Strobel, their imposition has been strenuously opposed.

Gary Strobel, a scientist at Montana State University, deliberately chose to defy existing regulations in order to continue his research on Dutch Elm Disease: his argument, that the present rules were "almost ludicrous."[35] Strobel was reprimanded by various federal agencies and forced to cut down his experimental trees and burn then in a university incinerator.[36] Although Strobel's acts were extreme, they reflect the prevailing scientific antipathy toward regulations and regulatory structures. Indeed, in 1987 the National Academy of Sciences issued an assessment of recombinant DNA guidelines not unlike Strobel's. Its argument was that genetically altered organisms were no more harmful than "a breeder's new variety of flower," and thus the regulations were superfluous.[37]

Scientists have sometimes threatened to leave their native countries if their work is forbidden. When recombinant DNA guidelines were initially introduced, for example, great care was taken by NIH to disseminate them throughout the world, in the hopes that a general parity would be maintained. Donald S. Frederikson, then director of the NIH, noted his fears: "had guidelines, of grossly uneven character sprung up within countries, or among them, the new biotechnology industry based on recombinant DNA methods conceivably would have risen in the Third World, or on ships 'beyond the 12 mile limit' in the fashion of gambling casinos."[38]

Similarly, "physician resistance" is a common complaint of medical ethics committees. "It is well known that many physicians are indifferent to and frequently hostile to what they regard as an 'intrusion' of ethics into their turf."[39] Physicians find ethics committees often time-consuming and sometimes threatening. Their own prognoses and treatment decisions may be challenged. To physicians, such committees often represent a mounting series of hurdles for difficult cases.

It is not surprising that physicians and scientists wish to see their work unhampered and the number of bureaucratic impediments to each stage of their progress minimized. This is a rational response to a heavily bureaucratized society. Moreover, it is also clear that much of the work of the various boards and committees would have been arrived at by practicing scientists and physicians themselves. Most safety precautions would have been taken anyway; most research proposals indicate a genuine respect for human subjects; most medical decisions reflect a physician's thoughtful assessment of the wishes of both the patient and family.

These structures, then, primarily serve to capture and address the unique case, the exception, that which may violate fundamental assumptions of safety or consent or where competing values and concerns are at stake. That these cases are few is not surprising. The very existence of such structures, as noted earlier, may itself be a deterrent to the submission of medical decisions or research projects that fail to take into account physical, social, or ethical guidelines. As such, these structures clearly serve a purpose. Though they often represent significant delays before medical or scientific work can progress, they serve as skeletal safety nets, which may weed out at least the worst abuses and identify the most complex questions.

Many of the regulatory forms and structures cited in this chapter were created *after* the disclosure of some major violation of rights or a major dilemma in decision making: i.e., biosafety committees after fears of biohazards, medical ethics committees after the Karen Quinlin and Baby Doe cases, IRBs after the Tuskegee revelations, one of the presidential commission reports at the request of theological groups. Their very structures were responsive to these problems. Thus they emerged as a form of problem solving, an effort to prevent abuses and bring several points of view to bear on critical issues. Moreover, in many of the committees, we find an attentiveness to social as well as scientific concerns, considerations of individual or social frailty and vulnerability. In this way, they may be thought of as implementations of the medical dictum *primum non nocere* ("first do no harm"). As David Thomasma argues, "an ethics committee may be personified to represent 'medicine' or 'social welfare' writ large."[40] This might be said of all these committee structures.

In such structures, we may in fact find the first steps toward a balancing of forces between Genetic Welfare and Social Welfare. Since scientific and technical issues are not solely paramount here, other considerations enter the arena of decision making. Social, ethical, and legal issues may play an equal role. The cost-benefit calculus may shift to some degree. The human element—considerations of suffering, injury, rights, and responsibilities—is brought more fully to the fore.

There is much work yet to be done on these structures. As they are presently constituted, they are too few in number, too diffuse in organization, and too varied in their roles to be of major utility. Yet their inclusion of the public, as members of the committees, may serve to balance the competing claims of progress in science with the need to prevent human suffering and social loss. If these structures begin to function to balance the quest for knowledge with protection of the public, most particularly its vulnerable members, they could begin to mitigate the more most serious effects of a swing toward Genetic Welfare.

CONCLUSIONS

The impetus to forward movement in the sciences is historic. The mystique of science, its promise and its potential for great works, has made its limitation a rare and reluctant imposition. In recent years, however, the abuses and failures of science—from Thalidomide to Tuskegee to the *Challenger*—have made the sciences more suspect, both by the public and the government. Moreover, the ethical dilemmas raised by medicine, over extending life as in the Karen Quinlin and Baby Doe cases and selecting patients for exceptional care, as in dialysis and organ transplantation, have made individual decision making a less than desirable situation.

There have arisen in recent years a variety of mechanisms to forestall abuses and to broaden the nature of decision making: biosafety committees, medical ethics committees, institutional review boards, and presidential commissions. These structures are imperfect. Biosafety committees have virtually withered away; medical ethics committees are diffusely structured; institutional review boards may be overlaid with bureaucratic demands; presidential commissions are often not clearly linked to decision making. Yet all are multidisciplinary and to some degree incorporate testimony and input from the public. As such, they are a first step in balancing the values of science with the interests and concerns of other groups. This emerging balance suggests that the *idea* of such structures, though not their current reality, is a direction we need to pursue in the future.

In the following chapter, we will consider the role of the public in defining medical and scientific problems. As in the creation of regulatory committees, the public has taken only a few tentative steps in seeking to modify the directions of genetic and reproductive technologies. But small though these steps may be, they are significant signposts for the directions we may expect in future years.

7

Citizen Participation and Public Policy

No sooner do you set foot upon American ground than you are stunned by a kind of tumult; a confused clamor is heard on every side, and a thousand simultaneous voices demand the satisfaction of their social wants. Everything is in motion around you; here the people of one quarter of a town are met to decide upon the building of a church; there the election of a representative is going on; a little farther, the delegates of a district are hastening to the town in order to consult upon some local improvements; in another place, the laborers of a villager quit their plows to deliberate upon the project of a road or a public school.

—ALEXIS DE TOCQUEVILLE,
Democracy in America

Citizen participation is an article of faith in the American vision of self-government. The right to organize, protest, and challenge is grounded in the very ethos of the American experiment. Since the New England town meetings, Americans have gathered in a vast array of settings and associations to debate public policy and to consider public issues.

Alexis de Tocqueville identified the American commitment to citizen action as a unique aspect of the American character, distinctive from that of the Europeans. He described the United States as a vast arena of activity, deeply responsive to public questions and intensely serious about the duties of civic participation.

Scientific and medical issues have not been exempt from such citizen involvement. In recent decades, several protest and social movements have emerged over scientific and medical issues. Some have questioned the location of research laboratories; others have challenged the directions of scientific and medical policy. These actions constitute important new developments in the history of citizen participation. This chapter will examine these public initiatives and consider their sources, evolution, and impact.

Of specific interest to us will be the assumptions and arguments forged by the public. Does the public support medicine and science? On what grounds? What are the public's fears? What are the expectations? To what degree can the public contribute to the agenda of science and medicine? Can the public alter this agenda to a significant extent? To begin our analysis, we will consider several contrasting views on these questions.

CITIZEN PARTICIPATION AND THE PUBLIC AGENDA

Can citizen participation help define the public agenda? This is a matter of some dispute. Pluralists, deriving their views from de Tocqueville, see citizen action as the centerpiece of public policy. From this perspective, shifting coalitions of interest groups routinely emerge to raise social issues and seek social action. Ultimately, to pluralists, a complex mosaic of power axes is formed, with no single group determining the critical decisions of society.

Alternatively, theorists who take a power-elite or elitist view of politics see a small number of powerfully situated groups setting the public agenda. From this perspective, citizen action is viewed as peripheral to true decision making. Citizen actions may address issues already on the agenda, but they do not pose the important questions or alter public policy to any meaningful extent.[1] Here, citizen action is to some degree viewed as window dressing.

Those who see citizen action as mere ornamentation often raise the issue of "co-optation." They argue that conflict is often muffled by giving protesting citizens formal roles within existing bureaucratic structures. By then offering the rewards of recognition or status, the opposition may be bought off, the public co-opted, and citizen action immobilized. According to Peter Bachrach and Morton Baratz, for example, "a particularly potent form of cooptation is participatory democracy. In Philip Selznick's words, it gives the opposition the illusion of a voice without the voice itself and so stifles opposition without having to alter policy in the least."[2]

As noted in Chapter 5, this book takes a power-elite perspective, arguing that the existing structure is at best difficult to dislodge. Co-optation presents a constant danger, and meaningful citizen action is assumed to face extraordinary odds. Further, those seeking to change the direction of social and political life often have limited resources and few points of leverage. Real social change is rare.

Nevertheless, there have been a few successes, small and fragile though these accomplishments may be. The civil rights movement, the women's rights movement, and the disability rights movement have, at minimum, brought some issues to the fore and created new sensitivities. The Civil Rights Act, the National Rehabilitation Act, and the Americans with Disabilities Act are but a few examples of progress achieved. Social movements concerning technology also have had some impact, particularly in their efforts to stop nuclear power plants and raise environmental concerns. Although these actions may not be agenda setting, in the deepest sense of the term, they may modify social conditions for significant numbers of people.

The question addressed in this chapter is whether citizen movements can affect the direction of genetic and reproductive research. Here, because the subject matter is arcane, citizen action is difficult and the path encumbered. Unlike demands for rights, equity, or access, scientific and medical advances are not self-evidently public issues. It is only typically where public protection or public safety is involved that public issues emerge clearly. Social impact concerns are far less evident as public issues. These are difficult to "rally round" since they have little immediacy or focal point.

In the coming years, we will need to consider what role the public can play in addressing the problems of medicine and science. Should it be consultative and advisory? Should it be adversarial? Thus far, it has taken both forms, and we need to consider the relative successes and failures of each. Many of these actions have paralleled those of scientists and physicians. For example, soon after the Asilomar conference of scientists, public protests arose over the construction of recombinant DNA research facilities.[3] The most important and controversial of these took place in Cambridge, Massachusetts. In the following section we will explore this controversy and its implications for public policy.

THE CAMBRIDGE CONTROVERSY

> *Decisions regarding the appropriate course between the risks and benefits of potentially dangerous scientific inquiry must not be adjudicated within the inner circles of the scientific establishment. Moreover, the public's awareness of scientific results that have an important impact on society should not depend on crisis situations.*
>
> —CAMBRIDGE EXPERIMENTAL REVIEW BOARD,
> Guidelines for the Use of Recombinant DNA Molecule Technology in the City of Cambridge, *January 5, 1977*[4]

One of the earliest citizen protest movements to develop over genetic engineering took place in Cambridge, Massachusetts in 1976.[5] A joint effort by citizens and local public officials sought to halt the planned construction of a moderate risk (P-3) recombinant DNA research laboratory at Harvard University.

Cambridge Mayor Alfred E. Vellucci, after consulting with several scientists, convened the City Council to debate construction of the planned laboratory. The City Council, with the sanction of the mayor, then voted 9–0 to hold public hearings on the laboratory, public safety being the primary concern. As the mayor argued, "We want to be damned sure the people of Cambridge won't be affected by anything that would crawl out of that laboratory."[6]

The mayor represented himself, throughout the proceedings, as a spokesperson for the people, a protector of public safety. To emphasize this point, he opened the proceedings with a rendition of Woodie Guthrie's "This Land Is Your Land," sung by the Cambridge Public High School choir.[7] The mayor then urged the City Council to ban all recombinant DNA research for two years. The mayor's request failed. Instead, a resolution was passed by the City Council calling for a three-month "good faith" moratorium, later extended for another three months.[8] At the same time, the council voted to set up a citizen board to review the issue of moderate risk research.

This newly created citizen board, entitled the Cambridge Experimental Review Board (CERB), was composed of "a housewife, a student, a security guard, a pharmacist, a college administrator, a welfare mother, a rubbish disposal contractor, a court clerk, and a tax collector."[9] Their role was to serve as a grass-roots sounding board. They heard "some seventy-five hours of testimony" from scientists for and against the construction of the laboratory.[10]

The key issue under consideration, for the citizen board, was the assessment of risk. On this matter, they came to the conclusion that there could be no full assurance of safety, but only a prediction of *reasonable* risk. Ultimately, the board voted to accept what they viewed as a reasonable risk, as defined by the scientists, under somewhat strengthened conditions of safety management.[11] Thus, under considerable pressure from the mayor and in the glare of publicity, the citizen board did not choose to ban the research entirely. Instead, their decision was similar to that which scientists took themselves in the years following Asilomar. They choose to proceed, but with caution.

The report of the Cambridge Experimental Review Board met a mixed response. Some called it "sober, thoughtful, and conscientious," but complained "it did not go far enough"; others were pleased.[12] Most argued positively, for both the process and the outcome. David Clem, one of the participants in the committee, assessed the event this way: "What is the lesson to be learned from Cambridge? It is simple, straightforward, and I believe quite sound. The lay public has a right to be involved in decisions affecting their general health and safety, and, given adequate information and meaningful power, the lay public can address complex issues and resolve them equitably."[13]

How do we evaluate the impact of this event on the public agenda? Did it change public policy? Did it alter the public agenda to any degree? It is interesting to note that despite its somewhat conflictual beginnings, the ultimate outcome of the Cambridge action was one of caution, moderation, and an endorsement to proceed. This approach then set a precedent for similar hearings around the nation. In city after city, lengthy scientific testimony was introduced, then carefully weighed and evaluated, but only a very few alterations were made from the original plans.[14]

Was the citizen board co-opted? They were certainly brought within the formal process and given an important formal role. Yet their cautious decision making was perhaps as much a reflection of technical difficulties and obstructions of language, as it was of co-optation. For example, when the hearings began, the mayor told the scientists: "Most of us in this room, including myself, are lay people. We don't understand your alphabet, so you will spell it out for us so we will know exactly what you are talking about because we are here to listen."[15] Yet excerpts from the testimony suggest that the scientists only party responded to this request. Much of the testimony was still in the obscure language of science, difficult to follow, and technical in nature. Such communication difficulties suggest a critical need to better translate the risks and benefits of science into a more accessible language.

Without this, the public often has little choice but to accept what scientists tell them.

Overall, the Cambridge debate established the principle of public participation and incorporated greater safety protection into the process. But the hearings did not halt recombinant DNA research or change the direction of public policy. The scientific agenda was unaltered. Although scientists were fearful, when the public debate arose, that heavy restrictions might be placed on their research, no major changes were imposed.

The Cambridge controversy did not end with these hearings. A second debate took place in 1980, four years after the initial controversy. At issue was the establishment of a research and development genetic engineering firm, Biogen. In the following section we will consider the impact of this second controversy.

THE SECOND CAMBRIDGE CONTROVERSY

By 1980, the city of Cambridge had a standing citizen committee to oversee recombinant DNA research within its boundaries, the Cambridge Biohazards Committee (CBC). When plans for the establishment of a genetic engineering firm, Biogen, became known, CBC called a hearing to consider its application to locate in Cambridge. As in the first controversy, the new hearing was open to the public. "Unlike the first rDNA debate, public opposition was mild. No biologist testified against siting the new biotechnology facility or spoke in support of additional controls."[16]

In order to consider guidelines for the establishment and management of commercial firms, the city manager recalled the Cambridge Experimental Review Board (CERB), which had been created during the first controversy. The CERB then chose to meet jointly with the new CBC. "The joint committee developed a consultative relationship with representatives of Biogen, Harvard, and MIT. After several months of hearings and deliberations, the CERB-CBC review panel issued recommendations emphasizing safeguards against the promiscuous release of genetically modified organisms and to somewhat lesser degree, against occupational hazards. The Cambridge City Council voted their recommendations into law on April 23, 1981."[17]

The outcome of this hearing was both professional and highly rationalized. It was in fact a quiet and virtually routinized process. According to the Office of Technology Assessment, "in contrast to the extensive publicity surrounding the passage of the first rDNA law, this new enactment was accompanied by little public discussion, and was only mildly acknowledged by the national media."[18] Thus, in the first hearing much controversy attended the event, although little was changed. By the time of the second debate, even the controversy was gone.

It is important to consider the changes that occurred in four years. First, as we have seen in the preceding chapter, scientific concerns over the dangers of recombinant DNA had waned by this time. Regulations were withering and a consensus had been reached, by the scientific community, that scientists could by and large regulate themselves.

Another reason for the relatively low profile of this second hearing may have been its consultative relationship with the professional organizations involved. The hearings were conducted within traditional and highly formalized bureaucratic channels. As in the first controversy, citizens were brought "inside" a decision-making structure. This may indeed be co-optation.

Perhaps the most important reason for the decline of the controversy was that the siting of the commercial firm simply became "business as usual." As argued in Chapter 2, this process is occurring rapidly, spurred by the commercialization of new technologies. The acceptance of Biogen suggests how quickly genetic engineering can become "institutionalized" in a community. What initially was seen as threatening became in the course of time expected and commonplace. More significantly, the public response to genetic engineering became casual and less concerned.

The inclusion of the public in oversight committees, thus far, does not appear to be a force for significant social change. The potential for co-optation, the routinization of the process of approval, the technical nature of the problems, and the towering expertise of scientists tend to push outcomes toward their original goals. As with the public's role in biosafety committees, they often serve as "rubber stamps." Thus, although some modifications may be achieved, toward greater public safety, to this date, the public has been fundamentally unwilling or unable to alter the direction of scientific research.

Thus far, we have examined citizen actions that have engaged the public but have had little real impact on pubic policy or the public agenda. We turn now to another form of citizen participation, the Oregon Health Decisions Project, which has also attempted to have an impact on public policy, this time in the field of health care. This movement, unlike the Cambridge controversies, is a long-term planning effort, seeking to substantially redesign the health care agenda.

THE OREGON HEALTH DECISIONS PROJECT

> *A new movement is spreading rapidly across this country that rejects the view that the tough ethical decisions belong solely in the province of the "experts"—it asserts that society must decide, and must do so by consensus.*
>
> —ANDREW BURNESS,
> "Who Lives, Who Dies, Who Pays?"[19]

In medicine, as in science, recent public efforts have sought to offset the decision making of "professionals" with more inclusive or democratic structures. The Oregon Health Decisions Project is the most significant of these new initiatives. It has become the catalyst for similar projects in more than fifteen states.[20] Each of these projects has sought to tailor medical care to the wishes and concerns of the public and to alter the existing direction of medical policy.

The Oregon Health Decisions Project was begun in 1982, its goal, to "give regular folks an organized forum in which to air their views on medical ethics issues."[21] Citizens from across

Oregon were convened to recommend policy for a broad range of issues, including preventive health care, death with dignity, adequacy of care, cost control, and the allocation of scarce medical resources.[22]

The Oregon Health Decisions Project created a multitiered structure of citizen participation. Small-group meetings were the initial phase. Here, trained volunteers led discussion sessions on bioethical and medical care issues, in "churches, community centers, [and] service clubs," using existing community networks as much as possible.[23] In Oregon, approximately 300 such meetings were held, involving more than 5,000 citizens. Town-hall meetings were the next phase, "intended to be a culmination of the hundreds of small-group sessions."[24] The results were designed to feed into an overall planning approach to Oregon's health care system.[25]

In 1984 the Oregon Project created a third phase, a "Citizens Health Care Parliament," patterned on the idea of a constitutional convention.[26] Representatives from the smaller town meetings were sent to the convention to draw up a statement of principles and priorities for health care. Thirty-four ranked principles were delineated by the convention, and the document became a working paper for future planning.[27]

Much of the effort of the Oregon Project was an attempt to redirect the medical agenda toward new priorities, primarily toward public health concerns and away from high technology orientations. Thus participants sought to discourage many forms of extraordinary care and to replace them with broader health goals. They argued: "The state should accept a leadership role to assure basic health care for all its citizens. Basic health care for all citizens is a higher priority than expending extraordinary resources for care which will benefit only a few individuals."[28] Similarly, Oregon's "Guide for Community Action and Ethical Decisions" noted:

> We have created a system which spends most of its resources on treating illness rather than promoting health; which is tilted heavily toward care provided in the most expensive hospital settings;

> which all too often rushes to accept new technologies without carefully assessing their worth; which has made death and dying an implacable enemy to be fought at all costs instead of a natural companion to life; which more and more is replacing humanitarianism with the profit motive as a guide to daily action.[29]

Public consensus on this issue led to action. "In 1987, the Oregon State legislature made an explicit policy decision to direct limited state Medicaid funds away from costly heart and liver transplants in favor of prenatal and other forms of preventive health care. Oregon Health Decisions played an instrumental role in assuring that citizen opinion and community preference shaped legislative opinion in the formulation of that decision."[30] Such a decision is controversial and reflects an appalling form of triage or brokering between human needs when applied to the poor. If conducted more broadly, however, these actions could reflect the wishes of the Oregon movement to implement a commitment to basic health care over extraordinary care in the face of limited resources.

A similar manifestation of the movement's resistance to high technology was in its concern for death with dignity. Participants repeatedly rejected highly invasive and extended treatments, and sought instead to reorient medicine toward greater autonomy for patients and families. They opposed the lengthy prolongation of life in favor of the enforcement of living wills, hospices, and the restraint of extraordinary care.

Overall, these actions led the Oregon movement toward an assessment of social as well as individual needs.[31] "Society must decide" begins Oregon's "Guide for Community Action and Ethical Decisions." Who defines "society" and how to reach "social decisions" was the question the Oregon Project raised in its initial deliberations.[32] This social vision is evident in the questions addressed by the Oregon Project. "Exactly what is 'adequate care'? What actions should be taken, if any, against the hospital that refuses to treat a non-paying patient?"[33] "Who should decide when the use of life-sustaining medical treatment should be foregone? On what basis should that decision be made? What balance should be struck between preventive medicine, acute care, and chronic, long-term care in what we as a nation spend on health?"[34] These are critical questions. They may indeed set the agenda for health care policy in the future.

The Oregon Health Decisions Project sees itself as the impetus for a "social movement" in health care.[35] Members have worked diligently to spread their ideas throughout the nation, and with considerable success. As various states have adopted its message, each has contributed a unique orientation. Maine, for example, has chosen to concentrate only on "the health care needs and services for the elderly."[36] Hawaii has used its ethnic organizations as a foundation for communication and community building.[37] Other states have concentrated on conferences, citizen education, or surveys of community attitudes. Yet the concern for a citizen voice, in creating a more equitable health system, has permeated all these efforts.

Unlike the Cambridge controversies, the health decisions projects have remained primarily citizens movements. Although they have been funded by private foundations, they do not appear to be dominated by experts. Their judgments appear somewhat freer and their challenges to the existing system somewhat broader.

Thus far, the health decisions projects have only peripherally addressed genetic and reproductive engineering issues, but the principles they have enunciated are relevant, their methods instructive, and their views of technology useful to consider. Their emphasis on preventive care, less technical approaches, and broad access and equity ultimately may serve as counterpoints to the highly technological tenets of Genetic Welfare. What they view as the social aspects of medicine could be particularly significant in raising the problems of vulnerable groups. It will be an important subject for future investigation to observe the orientations that emerge toward genetic and reproductive technologies, as they are taken up by the health decisions movement. They have the potential to meaningfully involve the public in weighing the directions of public policy and modifying the public agenda.

Overall, what has been emerging, through the types of movements described, is a potentially new role for the public in the determination of scientific and medical policy. Thus far, the impact has been slight and the agenda only minimally changed. Nevertheless, the role of the public is increasingly important, if only for the threat of change it represents. Because of this changing impact, public opinion on scientific and medical matters is being increasingly monitored. In the next section, we will examine some of the preliminary assessments of public opinion in these areas.

PUBLIC OPINION OF SCIENCE AND MEDICINE: A QUESTION OF TRUST

> *The public perception of science, by and large, still portrays the contemporary scientist as a selfless discoverer of truth, despite efforts on the part of sociologists and the media to show otherwise.*
>
> —SHELDON KRIMSKY,
> "The Corporate Capture of Academic Science and Its Social Costs"[38]

Does the public trust the practitioners of science and medicine? Does the public have confidence in their work? There is an emerging body of data on this question that indicates a slight but significant erosion in public trust. This erosion is most evident among the less educated and the lower class, groups that see themselves as vulnerable to technological change. In this section we will look at the nature of this development and its implications for Genetic Welfare.

In 1974 sociologists Amatai Etzioni and Clyde Nunn examined the evidence up to that date regarding the public assessment of science. Their conclusion was that from 1966 to 1973, a "middling shift" had occurred in the public's confidence in science, "not from great enthusiasm to great hostility, but from 'great confidence' to 'only some confidence.'" Etzioni and Nunn argued "that public approval of scientists did decline to a degree neither trivial nor monumental, neither reassuring nor alarming."[39]

Etzioni and Nunn attribute much of this shift in the public assessment of science to a "general disaffection from authority" that occurred during this period. Indeed, science actually rose in public ranking, relative to many other social institutions such as education, religion, labor, and the press. In 1973 only medicine outranked science in public support.[40]

More recent data support and extend Etzioni and Nunn's findings. Studies by the National Science Board in the late 1970s and early 1980s suggest a small negative shift, but with majorities of the public continuing their high positive assessments. According to John Walsh, "the American attitude toward science and technology continues to be decidedly favorable, although not as favorable as in the era of relatively uncritical approval in the 1950s."[41]

Similarly, the Office of Technology Assessment, in examining the National Science Board studies through the mid-1980s, found "in broad strokes . . . a public that has a high regard for the past achievements of science and technology and high hopes for even more spectacular results in the future."[42] At the same time, it also suggested "a level of wariness about some of the possible negative effects of science among a substantial minority of the American people."[43] Frequently stated apprehensions over science include a concern over the pace of change, the potential for a "few people to control our lives," a breakdown of people's ideas of right and wrong, and a loss of faith.[44] Overall, the Office of Technology Assessment found "some decline" in the numbers willing to acknowledge significant contributions to their lives from science, but found the majority of Americans still strongly supportive.[45]

While many assessments interpret these polls in strongly positive terms, some social scientists regard the general direction as more negative. Jack Desario and Stuart Langton argue that "between 1966 and 1976, the percent of the public expressing a great deal of confidence in the scientific community declined from 56 to 43 percent . . . and a 1979 survey indicated that 42 percent of the public believed that 'you can't trust what experts like scientists and technical people say because often what they say isn't right.'"[46] According to Desario and Langton, these figures suggest a "growing skepticism," leading to greater demands for citizen participation. They write, "as the public has become more uneasy about the impact of technology and the power of experts, citizens and many of their representatives have demanded greater participation in dealing with complex technological issues."[47]

Emphasizing this more negative analysis, it would also appear that the public does not feel fully informed about science or its methods. They see science as "remote" from their daily lives and difficult to clearly assess. This "remoteness," according to Etzioni and Nunn, may be gleaned from the number of "don't know" answers respondents provide about an institution. In their findings, "science received the *highest* percentage of 'don't knows,'" suggesting a sense of distance and a lack of information on the part of the public about this institution.[48] A similar sense of distance is reflected in the public's support for regulation. Interestingly, the public strongly endorses the idea of a government agency or external scientific body, responsible for making decisions about the uses and applications of genetic engineering technology.[49]

Who is most distrustful among the public? One finding, consistent across virtually all these studies of public opinion, is a positive correlation between education and income on the one hand, and attitudes toward science on the other. In general, the higher the education and income, the more confidence is expressed in science as an institution. The lower the education and knowledge, the greater the fear and apprehension.[50]

The low opinion of science, by those less educated and less monied, is a finding of considerable import. When institutions arouse such clear-cut class distinctions, there are underlying social reasons. The poor and the less educated apparently do not see the sciences contributing to their lives or well-being in the same way as do those of the middle and upper classes. They may also see themselves, in various ways, as victims of scientific development. The negative attitudes of the poor and less educated are important comments on the process of scientific communication and education to this portion of the public. It raises once again the need to consider the language of science and its ability to communicate to the public.

In support of these findings, recent studies have broken the public down into categories of "attentive" and "nonattentive" toward science. The "attentive" category, predictably, is highly educated, favorably disposed toward scientific endeavors, but small—only about 20 percent of the population.[51] Thus the largest and therefore more critical category of the public are the "nonattentives," those for whom science is a peripheral body of knowledge. This group is more skeptical of the benefits of science and more concerned about risk. Moreover, this group has manifested a consistent decline in confidence over the decades. According to John Walsh, the long-term implications of these findings are of concern. He notes that "between 1957 and 1979 the favorable majority declined from 87 percent to 66 percent, an erosion averaging 1 percent a year. Should that continue, science and technology would before too long face a deficit in public opinion."[52]

Despite these negative attitudes, some findings reflect the public's expectations that science can contribute to their lives. This result, consistent across many studies, is that science is typically evaluated in terms of its immediate contributions rather than long-term understandings. The National Science Board, for example, found that 81 percent of the public see science as making their lives "healthier, easier, and more comfortable."[53] Similarly, Etzioni and Nunn write, "large segments of the public do not clearly separate science from technology. People tend to see science as a set of practical tools that is part of technology."[54] This convergence of science and technology reflects several trends. First, as noted in earlier chapters, this perception corresponds in large part to a growing reality, because of the speed with which many scientific discoveries are now translated into applications. This view may also reflect a belief that the intents of science can be fundamentally positive, serving to enrich and enhance modern life. Such a perspective is, to some degree, an expression of appreciation for previous benefits received.

In general, although public opinion has become somewhat more distrustful of science and medicine, the fields have not wholly lost their luster. It would seem that despite apprehensions, the public continues to seek a contributing scientific establishment. Beyond this broad acceptance, however, there is also in the public attitude an expressed hesitancy to over-technicalize human life and a reluctance to leave decisions solely in the hands of scientists. This suggests that the public may function as a useful refractor for evaluating the emerging technologies and tenets of Genetic Welfare. While some may be embraced, others may be delayed, opposed, or challenged. In this way, the public has the potential to play a critical role in mediating the negative effects of genetic and reproductive technologies.

The increasing involvement in the public has startled the scientific and medical communities. In past eras, medical or scientific expertise automatically granted the prerogatives of autonomous decision-making power. This situation is now changing.

REACTIONS TO CITIZEN PARTICIPATION: RESISTANCE AND RESPONSIBILITY

> *Those [scientists] who came of age during the fifties and sixties may never quite understand why they have suddenly become "accountable" to a "participatory democracy."*
>
> —ROBERT MORRISON,
> "Commentary on 'The Boundaries of Scientific Freedom' "[55]

Overall, the scientific and medical responses to citizen movements have been wary. Like the professional reaction to regulatory committees, discussed in the preceding chapter, scientists and physicians are both surprised and alarmed over their newfound accountability. Certainly, as Robert Morrison has noted, traditional training and expectations have not led physicians or scientists to anticipate a "participatory democracy" in science.

In 1981 James D. Watson and John Tooze, early participants in the Asilomar Conference, wrote of their astonishment at the civic consequences of what they had originally seen as a professional action. They observed that "not even the most pessimistic (experienced?) of those initially involved raised the possibility that we would have to bring in the common man once we proposed . . . temporarily stopping certain experiments"[56] For most of those involved in Asilomar, the media attention and ensuing Cambridge struggles came as a considerable shock. Their altered status was not fully welcome.

To many scientists, citizen participation is antithetical to the very traditions of science. "Peer review," for example, often has been accepted as dialogue. Going beyond scientific boundaries evokes images of restriction and a loss of professional autonomy. Gerald Holton, for example, asks, "How can public participation be arranged without clashing with the very meaning of science as a consensual activity among trained researchers?"[57] The idea that scientists need only talk to other scientists is deeply ingrained. As Dorothy Nelkin observes, "Among scientists it is feared that widespread public involvement in decisions concerning science would virtually paralyze the conduct of research."[58] There is a strongly held fear, on the part of scientists, that the public will begin to pick and choose among scientific projects, perhaps favoring those with practical implications while downgrading those involving risk or expense.

Despite these misgivings, a number of professionals have entered into public dialogue. Scientists such as Edwin Chargaff, Ruth Hubbard, and George Wald, for example, helped initiate the Cambridge controversy. Their actions, perhaps more than any others, served to inform the public of a potential problem and spark the process of debate. Similarly physicians, psychiatrists, nurses, and social workers have been active consultants in the Oregon projects. Health care providers have given technical assistance, staff support, and organizational resources to the proceedings. They have also played a critical role in defining problem areas and identifying ethical dilemmas needing public attention.

The professional contributions to both Asilomar and the Oregon projects have been critical. As Daniel Callahan points out, "It is very hard, if not impossible, for the public to get interested in scientific decision making, unless potential social and ethical issues are called to their attention *by scientists*."[59] Scientists and physicians lend legitimacy to the causes of concern. Their involvement suggests that the issues are real and have been contemplated by "authorities."

Dorothy Nelkin argues that despite the claims of scientists, no absolute right exists to freedom of scientific inquiry. Instead, she suggests this right should be viewed as a "product of continued negotiations."[60] While perhaps cumbersome, this vision has great merit. In a society that has been alerted to so many recent dangers, the obligation to inform, compromise, and accommodate seems a fundamentally democratic enterprise. Scientists can no longer insulate themselves from the demands of public involvement. Moreover, their participation in a social dialogue serves an educative purpose. It may function to illustrate the commitment of scientists to their pursuits while encouraging citizens to become involved in important issues within their community.

CONCLUSIONS

One optimistic view of citizen participation, emerging in recent years, is that it may serve to offset many of the problems of our technocratic age. This perspective sees citizen participation as evolving, over time, from a reactive to a proactive stance. Citizen participation "has shown an evolution of techniques in which citizens first simply complained about policies and programs after they were implemented, then learned to react to nearly completed proposals, then demanded to take part in designing policies and programs, and finally became involved in designing sets of policies into alternative visions for the future of their communities."[61]

Whereas citizen participation may not be a panacea for all technological problems, the public does view issues in somewhat different terms than professionals and experts in their respective fields. The history of citizen participation, to date, has shown the public, overall, to be still deeply respectful toward science and medicine, but somewhat more cautious and more willing to regulate and to control. Moreover, it seems clear that citizens wish to have at least a consultative role in the formulation of public policy. This can be seen in the Cambridge debates and in public opinion generally.

The Oregon movement has taken a more pronouncedly different position from the traditional approaches of medicine. Alterations of emphasis and priority have emerged most prominently, reorganization to some degree. If this movement continues to spread, it is likely to have a considerable impact on the directions of medical policy.

The evidence also suggests that the public is most compliant when the language of science is most technical and the needed responses are relatively rapid, in situations of crisis or immediacy. In contrast, the Oregon movement, which is a long-term planning movement, has successfully conducted citizen information programs and generated a public language, intended to communicate the subtleties of the issues. This has permitted a more complex assessment of health care needs, with greater emphasis on planning and policy.

Perhaps the most successful medical change models to date have been outside the genetic arena, conducted by social activists, particularly women's groups. Through lobbies and lawsuits, selected medical products have been banned and certain sterilization practices better regulated. These small successes achieved by women's groups are evidence of the degree of militancy needed to actively alter the distribution of products, technologies and medical practice. Current concerns over genetic intervention have not reached this stage, although surrogacy may ultimately be a similar issue.

The examples presented in this chapter describe the first tentative steps the public has taken in seeking to alter the direction of medical and scientific policy. Thus far, only small changes have been achieved. Yet it would appear that with increasing participation and a substantial investment of time, the public can choose to encourage paths different from those of the professionals. Ultimately, negotiation or confrontation between scientists and the public may be needed. But public involvement is critical. Since it is the public who will ultimately bear the risks of much of the new technology, a public role is essential in helping to chart the directions science will take.

8

BALANCING THE SOCIAL AGENDA AND THE GENETIC AGENDA

Human beings have treated one another badly for as long as we have any historical evidence, but modernity has given us a capacity for destructiveness on a scale incomparably greater than in previous centuries. A social ecology is damaged not only by war, genocide, and repression. It is also damaged by the destruction of the subtle ties that bind human beings to one another, leaving them frightened and alone. It has been evident for some time that unless we begin to repair the damage to our social ecology, we will destroy ourselves long before natural ecological disaster has time to be realized.

—ROBERT BELLAH ET AL.,
Habits of the Heart: Individualism and Commitment in American Life

Two arguments dominate this book: one, that a vision of genetic intervention is coming to displace a social vision of the world; two, that this emerging worldview, Genetic Welfare, will create serious social costs for groups that are vulnerable—the stigmatized, the powerless, the alienated.

I have argued that an increasing reliance on genetic intervention may serve to subvert recent social gains made by disadvantaged groups. The stigmatized, just beginning to gain mainstream status, are newly labeled by genetic technologies. The powerless are further excluded from the benefits of society. The medically alienated are increasingly estranged by society's enlarging dependence on technological approaches. The organizations and movements that have sought to temper the impact of genetic intervention are significant, but thus far, lack major impact; they have served to modify the timing of interventions and increase safety. These are important accomplishments, yet restrained in their effects.

A final issue that needs to be addressed is how to frame a new public agenda. What policy is needed? What strategies should be pursued? What goals must be kept in view? As examples, we will briefly consider four illustrative issues: infant mortality, life expectancy, the doctor-patient relationship, and human selection. These subjects cross-sect the boundaries of both social and biological concerns; each is a policy dilemma that can be addressed through interventions that are social, biological, or a combination of the two.

In each case, the question of compelling need is considered. If genetic strategies come to predominate, what problems will be left unanswered? Whose gain is whose loss? Where should we invest limited resources? What is the needed balance between the social and the genetic?" This is not to suggest a simple quid pro quo; there is obviously no direct dollar exchange. Rather, it is a question of what commands our attention and captures our imagination. Our worldview affects the strategies we pursue. Whom do we see as most in need of help? Which solutions do we see as most beneficial? What are our plans for the future? These issues of priority and balance will be critical in the ensuing years.

INFANT MORTALITY

> *The infant mortality rate has been called "the most sensitive index of social welfare and of sanitary improvements which we possess."*
>
> —DENNIS WRONG,
> Population and Society[1]

One of the most important indexes of a society's investment in its human potential is its infant mortality rate. Infant mortality rates reflect the size and shape of a nation's social and medical institutions. The rate is sensitive to small changes in public policy and suggests the relative state of a society's well-being.

As a nation, the United States lags far behind most other industrialized nations of the world. According to the Children's Defense Fund, "The overall U.S. infant mortality rate in 1986 placed it eighteenth worldwide, behind Spain, Singapore, and Hong Kong. When considered alone, the white infant mortality rate placed the nation tenth worldwide; the black rate placed it *twenty-eighth,* behind Cuba, Bulgaria, and Czechoslovakia and tied with Hungary, Poland, and Costa Rica."[2]

Specific demographic trends are even bleaker. City data indicate appalling trends; many cities now have *rising* infant mortality rates, a phenomenon first observed in the 1980s. Further, the Children's Defense Fund notes, "A black infant born in Indianapolis was more likely to die in the first year of life than an infant born in North or South Korea. A black infant born in Boston, Chicago, Detroit, the District of Columbia, Indianapolis, Los Angeles, or Philadelphia was as likely to die as an infant born in Trinidad or Tobago."[3]

The most frequently noted cause of infant mortality is prematurity and low birth weight arising from a lack of prenatal care. A recent study conducted by the Alan Guttmacher Institute found that more than 550,000 American women "give birth each year without any health insurance. As a result, most have insufficient prenatal care, often leading to premature births and low birthweight babies, significant complications with the birth, and greater infant mortality."[4] Overall, 37 million Americans are without health insurance, including 9.5 million women between the ages of fifteen and forty-four. If we add in those who have private health insurance but no maternity coverage, 14 million women are without insurance for maternity care.[5]

Other statistics suggest the consequences of this situation. The Children's Defense Fund reported the following in 1989:

- 24 percent of pregnant women received no early prenatal care

- 32 percent of pregnant women received inadequate prenatal care
- 150,000 infants were born to mothers who received no prenatal care until the final three months of their pregnancy
- 70,000 infants were born with no prenatal care
- 4 million women of childbearing age, including 400,000 pregnant women and their fetuses, were exposed to excessive levels of lead.[6]

Thus, while vast sums of money are invested in genetic screening and reproductive techniques, we permit, at the same time, thousands of infants to be born each year with severe health problems that are *preventable*.[7] More bluntly, while we utilize expensive genetic procedures for middle- and upper-class women, to prevent diseases such as Down's syndrome, we permit lower-class women to give birth to underweight, undertreated, and chemically exposed babies, producing similarly injurious forms of disease and retardation. We invest huge sums of money in reproductive technologies such as in vitro fertilization while failing to invest in basic forms of health care. Ironically, the very first IVF clinic in America was established in Norfolk, Virginia, yet not many miles away, in Washington, D.C., is the highest infant mortality rate in the land, a rate close to double that of the national average and equivalent to that of many third world nations.[8]

The question we must ask, then, is what is of greatest benefit to society at this time, extensive investment in prenatal care or the continued expansion of high technology genetic and reproductive approaches? Is prenatal care as important to the health of society as the development of genetic screening technologies and reproductive technologies? Are lead enforcement programs as necessary as anmiocentesis or IVF or frozen embryo technologies? Would the full assurance of health care for all not serve the social good?

If we come to view genetic and reproductive intervention as the single or most important solution to the problem of infant disease and mortality, we will continue to generate a two-class society in health and well-being. Diseases, already disproportionate among the poor, will become rampant. The fabric of society, already taxed by inequities of class, will be further stressed.

There are many indications that, as a society, we are beginning to abandon the quest for greater social equality. The enlarging gap between the rich and poor, occurring in the 1980s, is but one example. Congressional data indicate that "taking inflation into account, the average family income of the poorest fifth of the population declined by 6.1 percent from 1979 to 1987 while the highest-paid Americans saw family income rise 11.1 percent."[9] These data indicate trends "inimical to the health of a democracy."[10]

A worldview that ennobles genetic approaches but ignores the social context will contribute to such inequality. Under present conditions of medical access, a strategy that relies on genetic intervention, to the detriment of social intervention, will worsen this trend. In such a situation, more expensive technologies will be pursued at the expense of more fundamental approaches. Those who cannot get even basic services will suffer.

Accessible prenatal care is not an untoward demand on society. It is so essential that it is difficult to accept the fact that so many women in this country are deprived of this most basic form of health care. Yet, for many, it is too expensive or at too far a distance to be accessible. This situation can be improved. Virtually all other industrialized societies do better than the United States.

A society is diminished by deaths that are the outcome of social negligence. Preventable losses are already too great in number. Too many times what might have been done, what we know how to do, has not been done. This is correctable; it is commitment that is at issue, not know-how. It is clearly reasonable to pursue genetic or reproductive strategies that will enhance our ability to prevent and treat disease. It is not reasonable to do so while currently available means of prevention and treatment are ignored. Data indicate that social intervention in these problems has a crucial impact. Class differentials speak clearly to this issue.

The visions of Genetic Welfare and Social Welfare both seek to address our contemporary problems. Both may be necessary to encompass the broadest constituencies and the fullest range of issues. Together they suggest the need for a two-fold strategy, one that addresses medical access, a second that incorporates high technology solutions. An approach that pursues both types of problems would embrace a far broader cross section of the population in a more democratic manner than we now achieve. But if we continue to invest in genetic interventions, while ignoring fundamental medical reforms, the health and well-being of this society will be severely damaged for generations to come.

LIFE EXPECTANCY AND THE HUMAN GENOME PROJECT

> *We used to think our fate was in our stars. Now we know, in large measure, our fate is in our genes.*
>
> —JAMES WATSON on the Human Genome Project[11]

The Human Genome Project is a new multibillion dollar research project seeking to decipher the DNA instructions that may define our molecular structure. Theoretically, this knowledge could help identify disease-causing genes and significantly enhance life expectancy.

The project has generated a high-flown rhetoric unusual in contemporary science. Harvard biologist and Nobel Laureate Walter Gilbert calls the undertaking the "holy grail of biology," which will usher in "the Golden Age of Molecular Medicine."[12] James Watson, one of the initial discoverers of the structure of DNA, declares, "I see an extraordinary potential for human betterment ahead of us. We can have at our disposal the ultimate tool for understanding ourselves at the molecular level. . . . The time to act is now."[13] George Cahill, vice president of the Howard Hughes Medical Institute, argues, "It's going to tell us everything. Evolution, disease, everything will be based on what's in that magnificent tape called DNA."[14]

The Human Genome Project is perhaps the quintessential expression of the idea of Genetic Welfare. It is a large-scale search for "human betterment," as James Watson argues, through genetic intervention. Powerful interests are in favor of the endeavor: Congress, which has allocated funds for the project, NIH, and many scientists. To some, the potential benefits appear enormous.

As in the case of infant mortality, however, we need to consider the paths we are taking, the decisions we are making, and the choices we are leaving behind. The paradox of a society that invests heavily in understanding the complexity of the human molecular structure yet fails to provide basic nutrition or adequate housing to a significant proportion of its citizens is something to ponder. At the very moment we are ushering in the "Golden Age of Molecular Medicine," the following conditions exist in the United States:

- 13 million children are living in poverty
- 500,000 children are estimated to go hungry each day
- 100,000 children are estimated to be homeless each night[15]
- 23 percent of all two-year-old children were not immunized for polio[16]
- 12 million children were exposed to toxic levels of lead paint.[17]

These conditions will not be solved by molecular genetics. Moreover, they will obviously account for a vast proportion of the disease, disability, and early death of the next several decades. Although cause and effect cannot be simplistically totaled up, it is reasonable to suspect that as much or more ill health could be prevented by addressing the above conditions as by decoding our DNA.

Why do we choose one path and not another? The deconstruction of DNA represents an intellectual puzzle of extraordinary proportions. It has been likened to "cartographers mapping the ancient world."[18] It is a challenge to our human abilities, a feat worth accomplishing. It is exciting science. Social welfare measures no longer seem to pose the same type of intellectual challenge. Today, they seem to many to offer little that is new, just a base line of acceptable treatment.

Americans, like most industrialized peoples of the world, have a fascination with the new, the different, and the breakthrough. These are often constructive affinities. Yet, as a nation, we need to consider whether we can afford to indulge a desire for intellectual challenge while there exists such a desperate need for basic care and treatment.

George Cahill has said of the Human Genome Project, "It will tell us everything." I think it will not. Unraveling our molecular structure may tell us something about the design of the human body, but not how to house a homeless family or feed a hungry child. Social well-being is more than the calculus of disease. "Human betterment" needs more than molecular analysis.

According to Mark Guyer of NIH's Human Genome Office, everyone will eventually "have access to a computer readout of their own genome, with an interpretation of their genetic strengths and weaknesses. At the very least this would enable them to adopt an appropriate life-style, choosing the proper diet, environment, and—if necessary—drugs to minimize the effects of genetic disorders."[19] Yet how many of us today can "choose" a proper environment? If clean air, purer water, and the elimination of toxic chemicals are not soon put on the public agenda, what environment will be worth choosing? How many cannot now, irrespective of their genetic readout, "choose" a proper diet? Are these prescribed drugs to be freely available? What of those with no health insurance? In short, these so-called choices make little sense in a context of limited options for so many.

There is, clearly, a reductionism in the attribution of social good to genetic structure. It reflects our society's need for neat solutions. As a nation, we are deeply troubled by the complexity of our contemporary problems. We see few paths to solving environmental pollution, homelessness, or the increasing impoverishment of women and children. It would be deeply reassuring to think we could solve many of these problems through genetic strategies. Thus, if we allow ourselves to believe that scientists are on the correct path, if what needs to be done has begun, if the "technological fix" is already "out there" waiting to be unleashed, then we need not bother with ungainly efforts to solve our social problems in other ways.

As a society, we urgently need to consider the interactive nature of our problems. Clearly, our exposure to chemicals, radiation, and pollution increase our propensity to the very diseases genetic technologies are supposed to "fix." Problems of poverty, homelessness, and malnutrition similarly increase the likelihood of illness. There is a frequently quoted metaphor in medical sociology which is useful to consider. It is the need to "focus upstream" a common problem in medical care. John B. McKinlay quotes Irving Kenneth Zola:

> "You know," he said, "sometimes it feels like this. There I am standing by the shore of a swiftly flowing river and I hear the cry of a drowning man. So, I jump into the river, put my arms around him, pull him to shore, and apply artificial respiration. Just when he begins to breathe, there is another cry for help. So I jump into the river, reach him, pull him to shore, apply artificial respiration, and then, just as he begins to breathe, another cry for help. So back in the river again, reaching, pulling, applying, breathing, and then another yell. Again and again, without end, goes the sequence. You know, I am so busy jumping in, pulling them to shore, applying artificial respiration, that I have *no* time to see who the hell is upstream pushing them all in."[20]

Genetic technologies are typically posed as "upstream" solutions, creating new paths to prevention. But social measures, too, are "upstream" efforts. Access to primary health care, environmental cleanup, and improvements in the quality of life also function to prevent ill health, disease, and disability; they serve the society as a whole. If we come to envision well-being in the broadest sense—genetic, social, environmental—our future agenda would be more substantial and its benefits far more broadly based.

THE DOCTOR-PATIENT RELATIONSHIP

> *The way a doctor listens to a patient, his ability to inspire the patient's confidence, to communicate that which must be communicated*

> *in a way that does not destroy hope are the things referred to as the "art of medicine." This is what medicine is all about, and this is what endures.*
>
> —NORMAN COUSINS,[21]
> *Foreword to* Medicine as a Human Experience *by David E. Reiser, M.D., and David H. Rosen, M.D.*

Not only is adequate medical care inaccessible to many citizens, but what exists is often presented in a form that is cold and calculating. Medicine is strong in its "hard" components—those aspects that are "equated with science." It is less than outstanding in its "soft" components, those aspects of medicine that include, among other considerations, the doctor-patient relationship.[22]

In the past several decades there have been many efforts to "humanize" the doctor-patient relationship, but these are tenuous accomplishments. For each new push for an increase in communicative skills, more personalized problem solving, or informed consent, there are many more introductions of technologies that newly distance the doctor and the patient.

The expansion of genetic and reproductive technologies, in particular, has exacerbated the problem of medical distance manyfold. New technologies often intervene directly between the doctor and patient. Even the definition of *patient* has changed. For example, in childbearing, where once the patient was the pregnant women, today, increasingly, it is the fetus. This has occurred because of our increasing technological ability to look inside the womb. Sonograms, for instance, according to Rosalind Petchesky, from their very first introduction, reduced the status of the woman and increased the significance of the fetus.

> I went back to trace the earliest appearance of these photos in popular literature and found it in the June '62 issue of *Look*. . . . It was a story publicizing a new book, *The First Nine Months of Life*, and it featured the now-standard sequel of pictures at one

> day, one week, forty-four days, seven weeks, etc. In every picture the foetus is solitary, dangling in the air (or in its sac) with nothing to connect it to any life-support system but a "clearly defined umbilical cord." . . . From their beginning, such photographs have represented the foetus as primary and autonomous, the woman as absent or peripheral.[23]

Such a technological sleight-of-hand alters our worldview. The woman becomes ever more diminished, the fetus enhanced.

A similar blurring of medical relationships is created by surrogate parenting. In this situation the dilemma becomes one of identifying the patient. Is it the biological mother, the biological father, the adoptive parents, the child? Whose "best interest" is central?

Genetic and reproductive technologies are, by necessity, transforming medical events into tacit or overt contractual relationships. Patients are expected to heed doctors' orders. If they fail to do so, they may be prosecuted for "prenatal abuse." Parents are expected to pursue all reasonable technological paths. If they fail to do so, they may be subject to "wrongful life" suits. Surrogacy entails an explicit contract, the violation of which has created several controversial court cases. All of this is then added to an institution that already practices "defensive medicine."

In short, the totality of technological change accumulates in a vision of medicine that is increasingly antagonistic and conflictual. Rather than moving toward a needed "ethicizing" of medicine and a greater attention to its more social aspects, we are heading instead toward a legalistic atmosphere. Both practitioners and patients are increasingly *liable* for their technological choices. We first permit and then promote the formulation of new expectations. More rapidly than we might expect, these expectations become institutionalized. The failure to meet these expectations then enters the arena of a legal or contractual battle.

When human relations are contractual, they are less humane, less personal, less spontaneous, and less comforting; they are more distant, more specified, more demanding, and more threatening. The end result is costly to the social fabric. The *gemeinschaft* nature of society grows more threadbare; the calculation of our social relations expands. Fundamentally, the technological imperative in genetic and reproductive medicine is moving us in the wrong direction in human relations. Rather than seeking to ameliorate a process so many observers agree is alienating, we are enhancing the very source of the problem.

Thus the benefits offered by new genetic and reproductive technologies come at a cost. While each new technology gives us additional opportunities to treat or cure a given set of problems, cumulatively, these technologies also alter the way in which we view one another. Technologies that permit us to overcome old obstacles give us new "vested interests." New opponents are created in the process: woman versus fetus, surrogate parent versus surrogate parent, parents versus wrongful lives, to name but a few.

How do we balance genetic and reproductive advance against a cost in social-medical relations? This is an obscure dilemma. No single technology functions to demean our human relations. Yet the cumulative effect of the expectations created by the range of new technologies is one that shifts our social relations toward a path of confrontation. A worldview that promotes genetic health as its ultimate goal and views disability as its ultimate burden fosters an adversary relationship.

As we embark on the use of new technologies, we will need to consider how they will alter the social relations of medical care. How will the technologies intervene between physician and patient? With what effect? At what cost? How can we soften the impact of technological advance?

The interplay between genetic and reproductive technologies and their social repercussions constitutes an important element in our future public agenda. The negative impact of these technologies can be modified if we are alert to their debilitating effects on human relations. If we are willing to commit our energies and attention to these issues at the same time as we continue to enhance our technological skills, we can compensate for their most problematic effects. But the tendency to distance the doctor and patient is likely to increase. It is critical that social dimensions as well as technological advances become a central focus of public attention.

HUMAN SELECTION

> *Eugenics . . . casts a shadow over all contemporary discourse concerning human genetic manipulation.*
>
> —DANIEL KEVLES, In the Name of Eugenics[24]

Genetic and reproductive technologies permit the application of selective criteria to human beings. As bioethicists often warn, we now have the means to determine "who shall live."

In the pursuit of good health, we have begun to tread a fine line in "human selection." We often choose to rule out certain diseases or, more accurately, certain human beings with those diseases. In some cases, as with Tay-Sachs disease, an as of now invariably fatal illness in early childhood, such a decision may be motivated by compassion. From many viewpoints, there is little quality of life in any sense traditionally understood, and great anguish and tragedy.

Other diseases, however, challenge our logic more severely; our sense of balance between cost and benefit is not as clear. Huntington's chorea is a case in point. Would a Woodie Guthrie be born today? Would his parents, as carriers of this disease, bear a child with the known risk? Could we now or soon screen him out prenatally? If the pace of genetic intervention continues, such an individual would not be born. Yet I, for one, am glad that he lived, although I mourn the anguish of his later life. One wonders, too, whether some perception of his coming illness contributed to the extraordinary creativity of his life.

Clearly, it is a just and meaningful desire to prevent fatal and debilitating diseases. Yet in pursuing this goal, we pay unobserved costs. In eliminating individuals with unwanted diseases, we also create a mind-set that justifies the process of human selection. We thus move into the questionable arena of human worth, and to some degree eugenic thought. We forego the idea of therapeutic change (i.e., dietary change or other forms of treatment) and opt instead for elimination. Individuals are seen as flawed. It is easier and more desirable to prevent their existence than work for their survival.

There is a eugenic risk in many of the reproductive technologies today. They permit us to pick and choose human beings or their characteristics, at least to some degree. The tenets of Genetic Welfare push us toward such decision making. The vision of "perfection" urges us to reject impaired aspects of humanhood. Newly available technologies make such decisions both possible and accessible. Financial pressures to limit the "burden" of care press us toward implementing these decisions.[25]

Many of the new technologies carry the "flavor" of eugenic thought. For example, according to Michelle Stanworth, "The first American physician . . . who assisted in establishing a 'surrogate pregnancy' justified his actions in the language of eugenics. 'I performed the insemination,' Dr. Simonds wrote, 'because there are enough unwanted children and children of poor genetic background in the world.'"[26] The pursuit of such eugenic improvement is not an unusual path.

It is important to consider the insidious nature of eugenic thought. There is a seductive element to eugenics, luring generation after generation to its call for "improving the race" and "bettering" the human being. Under different labels, it still has the same outcome; it tends to favor the selection of certain human beings, relative to others. The less powerful, the less perfect, are inevitably targeted.

In *Woman's Body, Woman's Rights*, Linda Gordon surveys the social history of reproductive movements over several generations. She notes a variety of shifts in emphasis, from the voluntary motherhood movement to the birth control movement to planned parenthood to population control. Interestingly, she finds that in every movement eugenic issues emerged, and in every movement they worked to undermine women's rights. Of the voluntary motherhood movement in the nineteenth century, for example, she writes, "Every eugenic argument was in the long run more effective in the hands of antifeminists than feminists. Motherhood was 'proved' to be weakened rather than improved, by higher education for women; those who argued that work outside the home devitalized motherhood had the best of the debate against those who supported job opportunities for women."[27] In a similar fashion, Gordon argues, eugenic issues emerged in the twentieth century, to the detriment of the poor, minorities, and Third World nations.

We need to balance the humane endeavor to rid the world of severe diseases with a concept of human dignity. The message of the various civil rights movements of the past thirty years, the black rights movements, the women's movement, the disability rights movement, the gay rights movement, is that we must respect human difference and offset frailty or past injustice with supportive services and advocate social arrangements. We need to retain these advances and not lose their small but beneficial contributions.

There is a vision in contemporary thought, albeit fragile, that endorses the concept of diversity. Although we do not always live up to this social and political ideal, the concept remains. Genetic Welfare, as a worldview, challenges not only the practice but our commitment to that idea. Genetic Welfare is a fundamentally undemocratic worldview. It discounts individuals on the basis of their inborn characteristics. Its vision is one of exclusion, not inclusion. It narrows the range of desirable human beings, identifying the different, the diseased, or the disabled as those we wish to "root out." This undermines the tenuous progress social movements have achieved in accepting human difference.

Alternatively, the ideals of Social Welfare promote social arrangements that are compensatory and mainstreaming. They acknowledge the legitimacy of human difference. The more widespread and routine supportive social arrangements are, the less stigmatizing will be their services and the more casually perceived their users. The greater the exposure, the less frightened we become. The continuum of what is accepted broadens.

Many years ago, writers Harold L. Wilensky and Charles Nathan Lebeaux defined institutional social welfare arrangements as those that are "normal, 'first line' functions of modern industrial society." In this view, social welfare "implies no stigma, no emergency, no 'abnormalcy.' Social welfare becomes accepted as a proper, legitimate function of modern industrial society. . . . The complexity of modern life is recognized. The inability of the individual to provide fully for himself, or to meet all his needs in family and work settings is considered a 'normal' condition."[28] Although it is not pleasant to think of economic tragedy and physical illness as normal and routine, they clearly are. Thus we need mechanisms that are systematically in place to acknowledge their existence, to administer humane treatment, and to promote social acceptance.

Once again, a balance must be sought between two imperatives, the one that seeks to prevent pain and suffering, the other that respects the rights of the less-than-perfect to participate in society. We will need to be vigilant in maintaining the continuity of mainstreaming and supportive services for those who become ill or disabled despite our best technological efforts. We will also need to be clearsighted in our commitment to diversity as well as to the prevention of disease.

WHERE DO WE GO FROM HERE?

> *The atomic age began with Hiroshima. After that no one needed to be convinced we had a problem. We are now entering the genetic age; I hope we do not need a similar demonstration.*
>
> —ROBERT SINSHEIMER,
> "An Evolutionary Perspective for Genetic Engineering"[29]

We clearly need to anticipate, to a greater degree than we now do, the broad impact of genetic and reproductive technologies. Their very existence will alter our world both in concrete and covert ways. With new abilities, we embark on new paths. These paths need to be monitored for their effects on those who administer them, those who use them, and, perhaps most importantly, on those who do not or cannot use them.

The introduction of such powerful technologies must alter, to use Robert Bellah's phrase, "the subtle ties that bind human beings to one another."[30] Relationships of power shift; our tolerance may be diminished; our alienation from the technological superstructure enhanced. We need, therefore, to be alert to these changes and to protect those who may be harmed. If we consider the gains created by the industrial revolution, we must also think of its victims; there were the early years of Dickensian conditions, the later years of sweatshops. All technological revolutions have their potential tragedies. But we can go far toward mitigating these costs if we choose to do so. Regulation, oversight, and planning are small but significant steps in this direction.

An interesting metaphor for social decision making in the use of technology lies in the small communal society of the Bruderhof. Unlike the Amish who reject technology outright, Bruderhof communities do use technology but select their items carefully.[31] For example, on a tour of one of their communities of some 400–500 people, I observed one television set, which they used to view major national events or issues of significance to their community, a televised hookup to connect them to their other communities, three cars which they shared for errands and trips, and woodworking machinery which was the basis of the community's income production. They had little else of a technological nature. Each of their tools was carefully chosen for its contribution to the community. The numbers in which they were accrued were similarly regulated. They *used* technology as a means to pursue their own ends, but were not subservient to it.

Contemporary society obviously cannot mimic the design of a small communal association. Yet the care with which the Bruderhof weighs and selects specific technologies carries an instructive element for the rest of us. The accumulation of technologies need not be beyond our control. Although technological development may continue, it need not dominate us. We can use it wisely.

The development of genetic and reproductive technologies will likely proceed at an accelerating pace. As with most other aspects of science and technology, we are not likely to go backward. Thus we must consider how to negotiate these developments so as to optimize their potential benefits for health and longevity while minimizing their negative consequences.

Traditionally, we have perceived science and medicine as the means to control nature, to give human beings power, so we are less subject to the whimsicality of unpredictable forces. In conquering many infectious diseases, for example, we transformed our world and our vision. But social policies, like science and medicine, are also forms of control; they permit us to pick and choose what dominates in the social arena, to buttress that which is favorable, and to mitigate that which seems to harm us. They too provide design and control.

In a modern complex society, systems cannot be left unregulated without serious abuses. From nursing homes to public utilities, from protection of the environment to the standards of education, we have found that some form of regulation is necessary to prod institutions to greater efficiency or to shield the public from harm.

It now seems clear that extensive regulation of genetic and reproductive engineering will be needed. Some minimum regulatory structures are already in place, such as those for recombinant DNA and research involving human subjects. Surrogate parenting is likely to be regulated in the future by most states.

At this time, however, these regulations are dispersed throughout our government. No single agency has the responsibility for consistent regulation or enforcement. But such regulation need not be ad hoc, falling between the gaps of federal, state, and local bodies. Clearly defined structures can be created; clearly allocated responsibilities can be defined. Ultimately, a Cabinet-level structure may be called for. Such a body could provide far greater oversight and protection than is now possible.

We must also consider the agenda of regulation. Regulation needs to be *broadly* construed to include a consideration of commitments and priorities as well as the more traditional bureaucratic requirements of safety and standards. What do we need most? Where do we begin? What is most urgent? Goals can be stated, unacceptable situations targeted.

Further, we need to examine the concept of early intervention. An instructive example is the atomic bomb. Since its creation and first use, the major nations of the world engaged in a continuing struggle to deter one another from using this dire technology. We negotiated and renegotiated treaties for decades, seeking to limit potential harm and control testing. The parallels between nuclear technology and genetic technology are cited by many. Thus we need to consider what lessons may be learned from the mistakes of the arms race and how long it has taken to assess its effects. If we foresee the potential abuses of certain technologies early, perhaps even prior to their appearance, we may be able to mitigate their harm. We need not be faced with another Hiroshima to act.

Such early interventions have precedent. The Food and Drug Administration is a case in point. They are mandated to intervene *prior* to the introduction of new drugs and treatments, to forestall potential abuses. Although the path toward approval is cumbersome, the outcome can be a better protected public. In the case of genetic interventions, a full range of issues could be considered, before their introduction. Social and economic questions as well as those of health and safety can be included.

Finally, the public must be involved. While this role has not yet been fully defined, the public can serve as "watch dog" for interests or emphases that are distinctive from those of professionals.[32] But, to make citizen involvement most meaningful, a shared language of science is needed, a means of communicating scientific ideas in clear and consistent form. The role of education in this process is vital. Social change at many levels will be needed, in the humanities as well as the sciences, to familiarize the next generation with this newly emerging terrain.

One of the most important contributions the public may offer in the fields of genetic and reproductive technologies is to monitor their speed of introduction. Technologies often outpace the social changes needed for their acceptance. The controversy over surrogate parenting is but one example. Public participation may lead to a slowing of use, a selectivity in making options available, and an insistence on legislative and regulatory structures, before their widespread introduction. Such planned social change should incorporate the needs of groups now excluded, including the poor and minorities. Public involvement of these groups would increase responsiveness to more diverse needs and broader constituencies.

In the next several decades, the contributions of genetic technology may be substantial. As scientists and physicians continue to unravel genetic puzzles, they may enhance the ability to predict, prevent, treat, or cure diseases. We, as a nation, may be swept up in a very deserved enthusiasm for such achievements. At the same time, however, we must continue to survey the social arena; to observe those who benefit and those who may be harmed. The needs of both populations must be kept clearly in view.

We must balance the two forces of Genetic Welfare and Social Welfare within our society. Both seek human betterment, but each contributes uniquely. The social agenda is clear, those in need apparent. Impoverishment, alienation, and discrimination are but a few of the essential problems. The genetic agenda is also clear: the identification, prevention, and treatment of genetic disease. If the ultimate goal is well-being, the pursuit of both social and genetic issues may serve the public good. But if we attach our sentiments, our resources, and our vision to the genetic agenda alone, while ignoring our social problems, we will harm the very foundations of our society. If, instead, we work to create a special sensitivity to those who suffer social injustice, the fragile threads that bind us will be strengthened.

Notes

INTRODUCTION

1. Throughout this book I will often group many of the new genetic and reproductive technologies. This is not to deny the unique effect of each, but rather to illustrate patterns of social influence and social change created by several technologies. Moreover, the technical line between the various technologies is blurring. For example, genetic screening permits the identification of genetic disease but may also be used to influence sex selection and genetic surgery. Similarly, in vitro fertilization is a reproductive technology but may be used in conjunction with genetic screening. The precedents for combining these various technologies under a single rubric are multiplying as their consequences increasingly overlap and intersect. Increasingly, writers now refer to these technologies as the "New Reproductive Technologies" (NRTs). See, for example, Patricia Spallone and Deborah Lynn Steinberg, eds., *Made to Order: The Myth of Reproductive and Genetic Progress* (New York: Pergamon Press, 1987); Gena Corea et. al., eds., *Man-Made Women: How New Reproductive Technologies Affect Women* (Bloomington and Indianapolis: Indiana University Press, 1987).
2. Much of the terminology used throughout this book, phrases such as *the biological revolution, reproductive technology,* and *genetic engineering,* are somewhat diffuse in meaning. This problem appears to be endemic to the field. For example, Daniel Callahan, director of the Hastings Center, notes that the term genetic engineering "has never had a strict technical meaning. . . . Loosely speaking, it has been used to refer to a wide range of possible and actual scientific breakthroughs—to in vitro fertilization and cloning . . . to genetic manipulation and recombinant DNA . . .

and to the development of new forms of human life. . . . On occasion, prenatal diagnosis is also included in the roster." See Daniel Callahan, "The Moral Career of Genetic Engineering," *Hastings Center Report* 9, no. 2 (April 1979):9. A similar difficulty exists even in drawing a distinction in these fields between science and technology. Norman Fost, for example, argues that "the distinction between science and technology is not always clear. The distinction blurs as the interval between basic and applied research narrows. Less than ten years elapsed between the discovery of recombinant DNA techniques and human applications, closely followed by profit-making uses." See Norman Fost, "Regulating Genetic Technology: Values in Conflict," in *Genetics and the Law, III*, ed. Aubrey Milunsky and George G. Annas (New York: Plenum Press, 1985), 16. Finally, a similar overlap exists between even science and medicine in these fields. The birth of Louise Joy Brown, for example, the world's first "test-tube baby," was both basic scientific research and medical application. See also Chapter 7 for further discussion of this issue.

CHAPTER 1. AN EMERGING WORLDVIEW

1. See, for example, John B. McKinlay and Sonja M. McKinlay, "Medical Measures and the Decline of Mortality," in *The Sociology of Health and Illness: Critical Perspectives*, 2nd edition, ed. Peter Conrad and Rochelle Kern (New York: St. Martin's Press, 1986), 10–23. As the McKinlays point out, there is considerable controversy over whether pubic health measures or immunizations were in fact responsible for the decline in infectious diseases. Whichever the most potent cause, the outcome was nevertheless a major and effective "assault" on the spread and severity of infectious diseases, leading to a lengthened life expectancy. See also Chapter 5.
2. Genetic screening includes such procedures as amniocentesis, fetoscopy, ultrasonography, alpha fetoprotein tests, and chorionic villus sampling.
3. Mitchell L. Zoler, "Genetic Tests Creating a Deluge of Dilemmas," *Medical World News* (September 22, 1986):35.
4. Verle Headings, "Optimizing the Performance of Human Genes," *The Humanist* (September/October 1972):10.
5. Alvin Weinberg, "Can Technology Replace Social Engineering?" in *Technology and Man's Future*, 2nd edition, ed. Albert H. Teich (New York: St. Martin's Press, 1977), 28.

6. James Danielli, "Industry, Society and Genetic Engineering," *Hastings Center Report* 2, no. 6 (December 1972):7.
7. See, for example, Charles Murray, *Losing Ground: American Social Policy, 1950–1980* (New York: Basic Books, 1984).
8. See, for example, Edward O. Wilson, *Sociobiology: The New Synthesis* (Cambridge, Mass.: Harvard University Press, 1975); Richard Dawkins, *The Selfish Gene* (New York: Oxford University Press, 1976); Arthur Jensen, *Bias in Mental Testing* (New York: Free Press, 1980); Hans Eysenck and Leon Kamin, *The Intelligence Controversy* (New York: Wiley, 1981); Richard Herrnstein, "I.Q.," *Atlantic Monthly* (September 1971):43–64.
9. See, for example, Stephen Jay Gould, *The Mismeasure of Man* (New York: Norton, 1981); Ruth Hubbard and Marian Lowe, eds., *Genes and Gender, II: Pitfalls in Research on Sex and Gender* (New York: Gordian Press, 1979); Richard Lewontin, Steven Rose, and Leon Kamin, *Not in Our Genes: Biology, Ideology and Human Nature* (New York: Pantheon, 1984).
10. See, for example, Alice Rossi, "A Biosocial Perspective on Parenting," *Daedalus* 106 (Spring 1977):1–31.
11. Edward O. Wilson, *Sociobiology*, 562. See also George W. Barlow and James Silverberg, eds., *Sociobiology: Beyond Nature/Nurture* (Boulder, Colo.: Westview Press, 1980); Ashley Montagu, ed., *Sociobiology Examined* (New York: Oxford University Press, 1980); "Will Biology Transform the Humanities?" *Hastings Center Report* 10, no. 6 (December 1980):27–39; Arthur L. Caplan, ed., *The Sociobiology Debate: Readings on Ethical and Scientific Issues* (New York: Harper and Row, 1978).
12. Lewontin, Rose, and Kamin, *Not in Our Genes*, 35.
13. William Ryan, *Blaming the Victim* (New York: Pantheon, 1971).
14. Cited in Richard Severo, "Federal Mandate for Gene Tests Disturbs U.S. Job Safety Official," *New York Times* (February 6, 1980):1.
15. Cited in Mary Sue Henifin and Ruth Hubbard, "Genetic Screening in the Workplace," *GeneWATCH* (November/December 1983):7.
16. See U.S. Congress, Office of Technology Assessment, *The Role of Genetic Testing in the Prevention of Occupational Disease* (Washington D.C.: Government Printing Office, 1983), Chapter 3. According to the Office of Technology Assessment seventeen U.S. companies conducted testing and fifty-nine companies were planning to or considering testing workers in the future.
17. Gina Kolata, "Genetic Screening Raises Questions for Employers and Insurers," *Science* 232, no. 4748 (April 18, 1986):318.

18. Ibid. See also Marc Lappe, *Broken Code: The Exploitation of DNA* (San Francisco: Sierra Club Books, 1984), Chapter 5; Beverly Merz, "Markers for Disease Genes Open New Era in Diagnostic Screening," *Journal of the American Medical Association* 254, no. 22 (December 13, 1985):3153–3161.
19. David E. Comings, "Presidential Address: The Genetics of Human Behavior—Lessons for Two Societies," *The American Journal of Human Genetics* 44, no. 4 (April 1989):457.
20. Ibid., 458, italics mine.
21. In the 1970s Sumner Twiss wrote extensively on this issue of parental responsibility. See, for example, his "Ethical Issues in Genetic Screening: Models of Genetic Responsibility," in *Ethical, Social, and Legal Dimensions of Screening for Human Genetic Disease*, ed. Daniel Bergsma (New York: National Foundation for the March of Dimes, 1974), 225–261; and "Parental Responsibility for Genetic Health," *Hastings Center Report* 4, no. 1 (February 1974):9–11. For a more current discussion, see Ruth Hubbard, "Fetal Rights and the New Eugenics," *Science for the People* (March/April 1984):7–29; and Ruth Hubbard, "Personal Courage is Not Enough: Some Hazards of Childbearing in the 1980s," in *Test-Tube Women: What Future for Motherhood?* ed. Rita Arditti, Renate Duelli Klein, and Shelley Minden (Boston: Pandora Press, 1984), 331–355.
22. Hermann J. Muller, "The Guidance of Human Evolution," *Perspectives in Biology and Medicine* III, no. 1 (Autumn 1959):15.
23. Bentley Glass, "Science: Endless Horizons or Golden Age?" *Science* 17, no. 3966 (January 8, 1971):28.
24. Werner Heim, "Moral and Legal Decisions in Reproductive and Genetic Engineering," in *Human Genetics and Social Problems*, ed. Thomas R. Mertens and Sandra Robinson (New York: MSS Corporation, 1973), 186.
25. See, for example, Barbara Katz Rothman, "The Meaning of Choice in Reproductive Technology," in *Test-Tube Women*, 23–33. See also Ruth Hubbard, who has written: "As 'choices' become available, they all too rapidly become compulsions to 'choose' the socially endorsed alternative. In this realm, it is amazing how quickly so-called options are transformed into obligations that, in fact, deprive us of choice." Ruth Hubbard, "Legal and Policy Implications of Recent Advances in Prenatal Diagnosis and Fetal Therapy," *Women's Rights Law Reporter* 7, no. 3 (Spring 1982):210.

26. President's Commission for the Study of Ethical Problems in Medicine and Biomedical and Behavioral Research, *Screening and Counseling for Genetic Conditions: The Ethical, Social and Legal Implications of Genetic Screening, Counseling and Education Programs* (Washington, D.C.: Government Printing Office, 1983), 54–55 [FN].
27. Ibid.
28. Ibid.
29. *Buck v. Bell*, Supreme Court of the United States (1927) 274 U.S. 200 47 S. Ct. 584, 71 L. Ed. 1000. Case presented in Thomas A. Shannon, *Law and Bioethics* (New York: Paulist Press, 1982), 340–341.
30. Alan Chase, in *The Legacy of Malthus: The Social Costs of the New Scientific Racism* (New York: Knopf, 1977), 16, estimates that approximately 63,000 people were sterilized between 1907 and 1964 under eugenic sterilization laws.
31. For a legal discussion of this issue, see *Reproductive Genetics and the Law*, ed. Sherman Elias and George J. Annas (Chicago: Year Book Medical Publishers, 1987), 162–167.
32. On the sterilization of welfare mothers, see *Relf v. Weinberger*, U.S. District Court for the District of Columbia (1974); on sterilization issues more generally, see, for example, Helen Rodriguez-Trias, "Sterilization Abuse," in *Science and Liberation*, ed. Rita Arditti, Pat Brennan, and Steve Cavrak (Boston: South End Press, 1980), 113–127.
33. Marc Lappe, "Moral Obligations and the Fallacies of 'Genetic Control,'" in *Ethics in Medicine*, ed. Stanley Joel Reiser et. al. (Cambridge, Mass.: M.I.T. Press, 1979), 360.
34. In 1975, the Committee to End Sterilization Abuse (CESA) was formed, later the Committee for Abortion Rights and Against Sterilization (CARASA). See Rodriguez-Trias, "Sterilization Abuse," and Adele Clark, "Subtle Forms of Sterilization Abuse: A Reproductive Rights Analysis," in *Test-Tube Women: What Future for Motherhood?* 188–212.
35. George Annas, "Pregnant Women as Fetal Containers," *Hastings Center Report* 16, no. 6 (December 1986):13; Marcia Chambers, "Are Fetal Rights Equal to Infant Rights?" *New York Times* (November 16, 1986):24.
36. John A. Robertson and Joseph D. Schulman, "Pregnancy and Prenatal Harm to Offspring: The Case of Mothers with PKU," *Hastings Center Report* 17, no. 4 (August/September 1987):26–27.

37. Thomas B. Mackenzie and Theodore C. Nagel, "When a Pregnant Woman Endangers Her Fetus," *Hastings Center Report* 16, no. 1 (February 1986):24–25.
38. Barbara Katz Rothman, "Commentary," *Hastings Center Report* 16, no. 1 (February 1986):25.
39. John C. Fletcher, "Attitudes Toward Defective Newborns," *Hastings Center Studies* 2, no. 1 (January 1974):21–32.
40. Theodosius Dobzhansky, "Man and Natural Selection," *American Scientist* 49 (1961):297.
41. Garrett Hardin, "The Moral Threat of Personal Medicine," in *Genetic Responsibility: On Choosing Our Children's Genes*, ed. Mack Lipkin, Jr., and Peter T. Ronley (New York: Plenum Press, 1974), 90, italics his.
42. Daniel Callahan, "The Meaning and Significance of Genetic Disease: Philosophical Perspectives," in *Ethical Issues in Human Genetics: Genetic Counseling and the Uses of Genetic Knowledge*, ed. Bruce Hilton and Daniel Callahan et. al. (New York: Plenum Press, 1973), 86.
43. Ibid.
44. Leon Kass, "New Beginnings in Life," in *The New Genetics and the Future of Man*, ed. Michael P. Hamilton (Grand Rapids, Mich.: William B. Eerdmans, 1972), 18.
45. Helga Kuhse and Peter Singer, *Should the Baby Live? The Problem of Handicapped Infants* (New York: Oxford University Press, 1985), Preface.
46. Kuhse and Singer deliberately speak of "killing" rather than euthanasia, since they see the two as fundamentally the same philosophically.
47. Thomas Kuhn, *The Structure of Scientific Revolutions*, 2nd edition (Chicago: University of Chicago Press, 1970), 11.
48. Craig Ellison, "The Ethics of Human Engineering," in *Modifying Man: Implications and Ethics*, ed. Craig Ellison (Washington, D.C.: University Press of American, 1977), 3.
49. Linda Bullard, "Killing Us Softly: Toward a Feminist Analysis of Genetic Engineering," in *Made to Order: The Myth of Reproductive and Genetic Progress* (New York: Pergamon Press, 1987), 110.

CHAPTER 2. A HIDDEN AGENDA

1. J. Robert Oppenheimer, *The Open Mind* (New York: Simon and Schuster, 1955), 88.

2. One unique exception to this is the abortion debate, which is distinctive in contemporary public life for its overt conflict of interest groups.
3. Arthur M. Schlesinger, *A Thousand Days* (Boston: Houghton Mifflin, 1965), 644–645.
4. Max Weber, *The Theory of Social and Economic Organization*, trans. A. M. Henderson and Talcott Parsons (New York: Oxford University Press, 1947), 340.
5. Jacques Ellul, *The Technological Society*, trans. John Wilkinson (New York: Vintage Books, 1964), 85–86.
6. Leon Kass, "The New Biology: What Price Relieving Man's Estate?" in *Bioethics*, ed. Thomas A. Shannon (New York: Paulist Press, 1976), 302.
7. Aaron Wildavsky, *The Politics of the Budgetary Process* (Boston: Little, Brown, 1964), 13.
8. Charles Lindblom, *The Policy-Making Process* (Englewood Cliffs, N.J.: Prentice Hall, 1968), 26.
9. Amy Virshup, "The Promise and the Peril of Genetic Testing: Perfect People," *New York* (July 27, 1987):31.
10. President's Commission for the Study of Ethical Problems in Medicine and Biomedical and Behavioral Research, *Screening and Counseling for Genetic Conditions: The Ethical, Social and Legal Implications of Genetic Screening, Counseling and Education Programs* (Washington, D.C.: Government Printing Office, 1983), 11. See also Richard Saltus, "Fetal Test Soon Could Be Routine," *Boston Globe* (July 29, 1985).
11. Cited in Sandra Blakeslee, "Genetic Discoveries Raise Painful Questions," *New York Times* (April 21, 1987):C1, C6.
12. See, for example, "All About the Baby," *Newsweek* (August 7, 1978):66; "The First Test-Tube Baby," *Time* (July 31, 1978):58; "The First Test-Tube Baby," *The New York Times* (July 26, 1978):1; "Test-Tube Baby Born—It's a Girl and It's Doing Fine," *Boston Globe* (July 26, 1978):1.
13. See, for example, "Vassar Brothers Agree to Offer Procedures for In Vitro Fertilization," *Poughkeepsie Journal* (September 30, 1987):4B.
14. Ibid.
15. Andrea Bonnicksen, "Embryo Freezing: Ethical Issues in the Clinical Setting," *Hastings Center Report* 18, no. 6 (December, 1988):26.
16. Donald K. Pickens, *Eugenics and the Progressives* (Nashville, Tenn: Vanderbilt University Press, 1968), 30.

17. Ibid.
18. Francis Galton, "The Possible Improvement of the Human Breed Under Existing Conditions of Law and Sentiment," *Nature* 64, no. 1670 (1901):664.
19. George Bernard Shaw, *Sociological Papers, 1904* (New York: Macmillan, 1905), as cited in Marc Haller, *Eugenics: Hereditarian Attitudes in American Thought* (New Brunswick, N.J.: Rutgers University Press, 1963):19.
20. Haller, *Eugenics*, 20.
21. On the confluence of politics, nationalism, and naturalism, see Pickens, *Eugenics and the Progressives*.
22. Kenneth Ludmerer, *Genetics and American Society: A Historical Appraisal* (Baltimore: Johns Hopkins University Press, 1972), 17–19.
23. Margaret Sanger, "Why Not Birth Control in America," *Birth Control Review* (May 1919):10–11. See also David Kennedy, *Birth Control in America: The Career of Margaret Sanger* (New Haven: Yale University Press, 1970).
24. This term was popularized by Theodore Roosevelt and the birth control movement. See, for example, Warren S. Thompson, "Race Suicide in America," *Birth Control Review* (August 1920):9.
25. For commentary on these two movements, see, for example, Jonas Robitcher, *Eugenic Sterilization* (Springfield, Ill.: Charles C. Thomas, 1973); John Higham, *Strangers in the Land: Patterns of American Nativism, 1860–1925* (New York: Atheneum, 1981); Daniel J. Kevles, *In the Name of Eugenics: Genetics and the Uses of Human Heredity* (New York: Knopf, 1985); Richard Hofstadter, *Social Darwinism in American Life*, revised edition (New York: Braziller, 1959).
26. Haller, *Eugenics*, 155–157.
27. For an elaboration of this idea, see Eric Hoffer, *The True Believer: Thoughts on the Nature of Mass Movements*, (New York: Harper and Row, 1951).
28. Other examples include gambling, alcoholism, and hyperactivity in children, to name a few. For an excellent discussion of the process of "medicalization," see: Renee Fox, "The Medicalization and Demedicalization of American Society," in *Doing Better and Feeling Worse*, ed. John Knowles (New York: Norton, 1977), 9–20.
29. Robert Francoeur, "We Can—We Must: Reflections on the Technological Imperative," *Theological Studies* 33 (1972):428;

See also Nicholas Crotty, "The Technological Imperative: Reflections on Reflections," *Theological Studies* 33 (1972):446–449.

30. Amatai Etzioni, *Genetic Fix* (New York: Macmillan, 1973), 99.
31. Barbara Ehrenreich and Deirdre English, *Complaints and Disorders: The Sexual Politics of Sickness*, Glass Mountain Pamphlet No. 2 (Old Westbury, N.Y.: Feminist Press, 1973), 5; see also Barbara Ehrenreich and Deirdre English, *For Her Own Good: 150 Years of the Expert's Advice to Women* (Garden City, N.Y.: Anchor Press, 1978); G. J. Barker-Benfield, *The Horrors of the Half Known Life* (New York: Harper and Row, 1976).
32. Ehrenreich and English, *Complaints and Disorders*, 17.
33. Richard W. Wertz and Dorothy C. Wertz, *Lying-In: A History of Childbirth in America* (New York: Free Press, 1977), 94–95.
34. Ibid., 93.
35. Cited in Mitchell L. Zoler, "Genetic Tests Creating a Deluge of Dilemmas," *Medical World News* (September 22, 1986):42.
36. See Marc Lappe, *Broken Code: The Exploitation of DNA* (San Francisco: Sierra Club Books, 1984).
37. In 1980, in the case of *Diamond, Commissioner of Patents, v. Chakrabarty*, the Supreme Court in a five-to-four decision held that the one-celled organism could be patented. For a discussion of this issue, see Stephen P. Stitch, "Genetic Engineering: How Should Science Be Controlled?" in *And Justice For All*, ed. Tom Regan and Donald Van DeVeer (Totowa, N.J.: Rowman and Allenheld, 1982), 86–115. It should also be noted that there is considerable controversy over whether these genetically altered organisms are in fact new life forms. This is itself a philosophical issue of considerable import.
38. Ellen Hale, "Putting Patent on Animals Creates Moral, Legal Furor," *Gannett News Service* (August 12, 1988).
39. Harold Green, "Chakrabarty: Tempest in a Teapot," *Hastings Center Report* 10, no. 5 (October, 1980):12.
40. A related practice has begun in terms of conceiving a child for treatment purposes (i.e., bone marrow transplants) although not necessarily by use of reproductive technologies. For further reading on fetal tissue transplant issues, see John A. Robertson, "Rights, Symbolism and Public Policy in Fetal Tissue Transplants," *Hastings Center Report* 18, no. 6 (December 1988):5–12; Kathleen Nolan, "Genug ist Genug: A Fetus Is Not A Kidney," *Hastings Center Report* 18, no. 6 (December 1988):13–19; Mary Mahowald, Jerry Silver, and Robert A. Ratcheson, "The Ethical

Options in Transplanting Fetal Tissue," *Hastings Center Report* 17, no. 1 (February 1987):9–15; Tamar Lewin, "Medical Use of Fetal Tissues Spurs New Abortion Debate," *New York Times* (August 16, 1987):1, 30.

41. Andrew Pollack, "Gene Splicing Payoff Is Near," *New York Times* (June 10, 1987):D1, D6.
42. See "Investors Dream of Genes," *Time* (October 20, 1980):72; "How Molecular Biology Is Spawning an Industry," *Newsweek* (March 17, 1980):17.
43. Subcommittee on Investigations and Oversight, Committee on Science and Technology, U.S. House of Representatives, *Hearings on Commercialization of Academic Biomedical Research* (Washington, D.C.: Government Printing Office, 1981), cited in Sheldon Krimsky, "The Corporate Capture of Academic Science and Its Social Costs," in *Genetics and the Law, III*, ed. Aubrey Milunsky and George G. Annas (New York; Plenum Press, 1985), 46.
44. Ibid., 52; see also Lappe, *Broken Code: The Exploitation of DNA*.
45. Diana B. Dutton, *Worse than the Disease: Pitfalls of Medical Progress* (New York: Cambridge University Press, 1988), 206.
46. Paul Starr, *The Social Transformation of American Medicine* (New York: Basic Books, 1982), 448.
47. Ibid.
48. Cited in Dutton, *Worse than the Disease*, 215.
49. Louis Knowles and Kenneth Prewitt, eds., *Institutional Racism in America* (Englewood Cliffs, N.J.: Prentice Hall, 1969).
50. Kristen Lucker, *Abortion and the Politics of Motherhood* (Berkeley: University of California Press, 1984), 158.
51. Peter Berger and Thomas Luckmann, *The Social Construction of Reality* (Garden City, N.Y.: Anchor Books, 1966), 91.
52. Ibid., 103.
53. Ibid., 112.
54. Francoeur, "We Can—We Must," 428.
55. Robert Redfield, *The Primitive World and its Transformations* (Ithaca, N.Y.: Cornell University Press, 1967), 92.

CHAPTER 3. THE PROBLEM OF STIGMA

1. It should be noted that deviance and disability are often treated together in sociological material, as subsets of nonnormative behavior. Although the issues are in many ways very different, the thread that links the two is their common position outside accepted or desired social interaction. Theoretical schools such as labeling

and stigma theory typically treat both deviance and disability as similarly labeled forms of behavior. This perspective will be adopted here. See, for example, Edwin Schur, *The Politics of Deviance: Stigma Contests and the Uses of Power* (Englewood Cliffs, N.J.: Prentice Hall, 1980); Vilhelm Aubert and Sheldon Messinger, "The Criminal and the Sick," in *Medical Men and their Work*, ed. Elliot Freidson and Judith Lorber (New York: Aldine-Atherton, 1972), 288–308; Nicholas Kittrie, *The Right to Be Different: Deviance and Enforced Therapy* (New York: Penguin, 1971).

2. "The Model Eugenical Sterilization Law" advanced by the Eugenics Records Office, upon which many state laws were based, included as "socially inadequate," and thus eligible for sterilization, the following categories: "(1) Feeble-minded; (2) Insane (including the psychopathic); (3) Criminalistic (including the delinquent and wayward); (4) Epileptic; (5) Inebriate (including drug habitues); (6) Diseased (including the turberculous, the syphilitics, the leprous and others with chronic, infectious, and legally segregable diseases); (7) Blind (including those with seriously impaired vision); (8) Deaf (including those with seriously impaired hearing); (9) Deformed (including the crippled); and (10) Dependent (including orphans, ne'er-do-wells, the homeless, tramps, and paupers)." Harry H. Laughlin, *Eugenical Sterilization in the United States* (Chicago: Psychopathic Laboratory of the Municipal Court of Chicago, 1922), 446–447.
3. See, for example, David Rothman, *The Discovery of the Asylum: Social Order and Disorder in the New Republic* (Boston: Little, Brown, 1971). Rothman notes that although a reform and rehabilitative model had been briefly attempted during the Jacksonian era, the approach was ultimately abandoned and rejected by the midnineteenth century. Until the twentieth century, mental hospitals, poorhouses, and penitentiaries served primarily as custodial warehouses.
4. Erving Goffman, *Stigma: Notes on the Management of Spoiled Identity* (Englewood Cliffs, N.J.: Prentice Hall, 1963).
5. Ibid., 3, italics mine.
6. Ibid., 5.
7. Howard Becker, *The Outsiders* (New York: Free Press, 1963), 9, italics his. See also Howard Becker, *The Other Side* (New York: Free Press, 1964); Kai Erikson, "Notes on the Sociology of Deviance," *Social Problems* 9, no. 4 (Spring 1962):308.
8. Schur, *The Politics of Deviance*, 9, italics his.

9. *Guidelines for Reporting and Writing About People with Disabilities* (Lawrence, Kans.: Research and Training Center on Independent Living, University of Kansas, 1984).
10. Jane Mercer, *Labeling the Mentally Retarded* (Berkeley: University of California Press, 1973), 1, 31, italics mine. See also Eliot Friedson, "Disability: A Social Deviance," in *Medical Men and Their Work*, 330–352; Stephen P. Spitzer and Norman K. Denzin, eds., *The Mental Patient: Studies in the Sociology of Deviance* (New York: McGraw-Hill, 1968).
11. See Adrienne Asch, who writes that all forms of disability are at least partly "socially constructed." Because of this, "disability's all-too-frequent consequences of isolation, deprivation, powerlessness, dependence, and low social status are far from inevitable and within society's power to change." Adrienne Asch, "Reproductive Technology and Disability," in *Reproductive Laws for the 1990s*, ed. Sherrill Cohen and Nadine Taub (Clifton, N.J.: Humana Press, 1989), 73.
12. Richard Scotch, *From Good Will to Civil Rights: Transforming Federal Disability Policy* (Philadelphia: Temple University Press, 1984); Steven A. Holmes, "House Approves Bill Establishing Broad Rights for Disabled People," *New York Times* (May 23, 1990); A1, A18.
13. Jeff Lyon, *Playing God in the Nursery* (New York: Norton, 1985), see especially 260–261; B. Ennis and L. Siegel, *The Rights of Mental Patients* (New York: Avon Books, 1973).
14. "Court: Airlines Free to Discriminate Against Disabled," *Poughkeepsie Journal* (June 28, 1986):1; See also "Most Airlines Held Exempt on Handicapped Rights Rule," *New York Times* (June 28, 1986):8.
15. Frank Tannenbaum, *Crime and Community* (Boston: Ginn, 1938), 19–20.
16. Marc Lappe, "Moral Obligations and the Fallacies of 'Genetic Control,'" in *Ethics in Medicine*, ed. Stanley Joel Reiser et. al. (Cambridge, Mass.: M.I.T. Press, 1979), 357.
17. James Sorenson, "Some Social and Psychological Issues in Genetic Screening: Public and Professional Adaptation to Biomedical Innovation," in *Ethical, Social and Legal Dimensions of Screening for Human Genetic Disease*, ed. Daniel Bergsma (New York: National Foundation for the March of Dimes, 1974), 174.
18. D. S. Halacy, *Genetic Revolution* (New York: Harper and Row, 1974), 118.

19. Cited in Mary Sue Henefin and Ruth Hubbard, "Genetic Screening in the Workplace," *GeneWATCH* (November/December 1983):8.
20. Sorenson, "Some Social and Psychological Issues in Genetic Screening," 174.
21. Patricia Jacobs, "Aggressive Behavior, Mental Subnormality and the XYY Male," *Nature* 208, no. 5017 (December 25, 1965):1351–1352; Ernest B. Hook, "Behavioral Implications of the Human XYY Genotype," *Science* 179, no. 4069 (January 12, 1973):139–150.
22. Jon Beckwith and Jonathan King, "The XYY Syndrome: A Dangerous Myth," *New Scientist* 64 (November 1974):474–476; "The XYY Controversy: Researching Violence and Genetics," conference sponsored by the Hastings Center, November 9–10, 1978/February 8–9, 1979, *Hastings Center Report* Special Supplement 10, no. 4 (August 1980):1–31.
23. Barbara Katz Rothman, "The Meaning of Choice in Reproductive Technology," in *Test-Tube Women: What Future for Motherhood?* ed. Rita Arditti, Renate Duelli Klein, and Shelley Minden (Boston: Pandora Press, 1984), 31.
24. Ibid.
25. In Laurie Nsiah-Jefferson, "Reproductive Laws, Women of Color and Low-Income Women," in *Reproductive Laws for the 1990s*, 33.
26. Cited in John Fletcher, "Genetics, Choice and Society," in *Genetic Responsibility: On Choosing Our Children's Genes*, ed. Mack Lipkin, Jr., and Peter T. Ronley (New York: Plenum Press, 1974), 95.
27. Ibid.
28. Linus Pauling, "Reflections on the New Biology," *UCLA Law Review* (February 1968):269.
29. "Counseling and Screening: Discussion," in *Genetic Responsibility*, 75.
30. Tabitha Powledge, "Genetic Screening," *Encyclopedia of Bioethics* (New York: Free Press, 1978), 569.
31. Dorothy Wilkinson, "For Whose Benefit? Politics and Sickle Cell," *Black Scholar* 5, no. 8 (1974):29; see also Mary Sue Henefin, Ruth Hubbard, and Judy Norsigian, "Prenatal Screening," in *Reproductive Laws for the 1990s*, 155–183 where this citation is reproduced.
32. "To Screen or Not to Screen for the Fragile X Syndrome," *Hastings Center Report* 17, no. 1 (February 1987):2.

33. Ibid.
34. Marc Lappe, "The New Technologies of Genetic Screening," *Hastings Center Report* 14, no. 5 (October 1984):19. See also Ruth Macklin, "Mapping the Human Genome: Problems of Privacy and Free Choice," in *Genetics and the Law, III*, ed. Aubrey Milunsky and George Annas (New York: Plenum Press, 1985), 107–114.
35. Milton Himmelfarb, "Biomedical Ethics and the Shadow of Nazism," conference sponsored by the Hastings Center, April 8, 1976, *Hastings Center Report* Special Supplement 6, no. 4 (August 1976):7.
36. Marsha Saxton, "Born and Unborn: The Implications of Reproductive Technologies for People with Disabilities," in *Test-Tube Women: What Future for Motherhood?* 306.
37. See, for example, Barbara Katz Rothman, *The Tentative Pregnancy* (New York: Viking, 1986); Madeline J. Goodman and Lenn E. Goodman, "The Overselling of Genetic Anxiety," *Hastings Center Report* 12, no. 5 (October 1982):20–27.
38. Leon Kass, "Making Babies—the New Biology and the 'Old' Morality," *The Public Interest* 26 (Winter 1972):48.
39. Marc Lappe, "How Much Do We Want to Know About the Unborn?" *Hastings Center Report* 3, no. 1 (February 1973):8–9.
40. Daniel Callahan, "The Meaning and Significance of Genetic Diseases: Philosophical Perspectives," in *Ethical Issues in Human Genetics: Genetic Counseling and the Uses of Genetic Knowledge*, ed. Bruce Hilton and Daniel Callahan et. al. (New York: Plenum Press, 1973), 86–87.
41. John Fletcher, "Attitudes Toward Defective Newborns," *Hastings Center Studies* 2, no. 1 (January 1974):21–32.
42. Ibid., 30.
43. Jean Harley Guillemin and Lynda Lyle Homstrom, *Mixed Blessings: Intensive Care for Newborns* (New York: Oxford University Press, 1986); Joseph E. Magnet and Eike-Henner Kluge, *Withholding Treatment from Defective Newborn Children* (Quebec: Brown Legal Publications, 1985); Robert Weir, *Selective Nontreatment for Handicapped Newborns* (New York: Oxford University Press, 1984).
44. Thomas Murray, "The Final Anticlimactic Rule on Baby Doe," *Hastings Center Report* 15, no. 3 (June 1985):5–9; "The Supreme Court and the Many Baby Doe Regulations," *Ethical Currents* 8 (August 1986):6.
45. See, for example, Adrienne Asch and Michelle Fine, *Women with*

Disabilities: Essays in Psychology, Culture and Politics (Philadelphia: Temple University Press, 1988); Mary Johnson, "Killing Babies Left and Right," *The Disability Rag* (Fall 1985):22–33; Marsha Saxton, "Born and Unborn: The Implications of Reproductive Technologies for People with Disabilities," in *Test-Tube Women*, 298–312; Anne Finger, "Claiming All of Our bodies: Reproductive Rights and Disability," in *Test-Tube Women*, 282–297.

46. See Adrienne Asch and Michelle Fine, "Shared Dreams: A Left Perspective on Disability Rights and Reproductive Rights," in *Women with Disabilities*, 297–305.
47. See, for example, Adrienne Asch, "On the Question of Baby Doe," *Health Pac Bulletin* (August 1986):6, 8–13.
48. Johnson, "Killing Babies Left and Right," 22–23; Finger, "Claiming All of Our Bodies," 282–297.
49. Emile Durkheim, *The Rules of Sociological Method* (Glencoe, Ill.: Free Press, 1938), 68–69.
50. Ibid., 68.
51. Ibid., 69.
52. Ibid., 71.

CHAPTER 4. THE PROBLEM OF POWERLESSNESS

1. The *Republic* has sometimes been called a eugenic tract. In the *Republic*, the following argument is made: "It follows from what we have just said that, if we are to keep our flock at the highest pitch of excellence, there should be many unions of the best of both sexes, and as few of the inferior as possible and that only the offspring of the better unions should be kept." *The Republic of Plato*, trans. Francis MacDonald Cornford (New York: Oxford University Press, 1945), 159.
2. C. Wright Mills, *The Power Elite* (New York: Oxford University Press, 1959); Floyd Hunter, *Community Power Structure* (Chapel Hill University of North Carolina Press, 1953); G. William Domhoff, *Who Rules America?* (Englewood Cliffs, N.J.: Spectrum, 1967); G. William Domhoff, *The Powers That Be: Processes of Ruling Class Domination in America* (New York: Vintage, 1979); Frances Fox Piven and Richard Cloward, *Regulating the Poor: The Function of Public Welfare* (New York: Pantheon, 1971); Frances Fox Piven and Richard Cloward, *Poor People's Movements: Why They Succeed, How They Fail* (New York: Pantheon, 1977).

3. The term *eutelegenesis* was coined by Herbert Brewer, a close associate of Hermann J. Muller. according to Daniel Kevles, "Brewer coined the word to refer to the eugenic breeding of human beings via pregnancies produced "from afar"—that is, by artificial insemination." Muller adopted the term and became a major advocate of the policy. See Daniel J. Kevles, *In the Name of Eugenics: Genetics and the Uses of Human Heredity* (New York: Knopf, 1985), 188.
4. Hermann J. Muller, "Human Evolution by Voluntary Choice of Germ Plasm," *Science* 134, no. 3480 (September 8, 1961):646.
5. Ibid., 646–647.
6. Hermann J. Muller, "Should We Weaken or Strengthen Our Genetic Heritage?" *Daedalus* 90 (1961):450.
7. Hermann J. Muller, "The Guidance of Human Evolution," *Perspectives in Biology and Medicine*, 3, no. 1 (Autumn 1959):35; William T. Vukovitch, "The Dawning of the Brave New World: Legal, Ethical and Social Issues of Eugenics," *Law Forum* 2 (1971):199.
8. Harold M. Schmeck, Jr., "Nobel Winner Says He Gave Sperm for Women to Bear Gifted Babies," *New York Times* (March 1, 1980):6; "Dr. Shockley's Genes," *New York Times* (March 4, 1980):14, Kevles, *In the Name of Eugenics*, 262–263.
9. Kirsten Aner, "Genetic Manipulation as a Political Issue," in *Genetics and the Quality of Life*, ed. Charles Birch and Paul Albrecht (Elmsford, N.Y.: Pergamon Press, 1975), 62.
10. Ibid., 63, emphasis hers.
11. "'Million-dollar Baby' in Good Health," *Knight-Ridder Newspapers* (June 6, 1986):3A.
12. On of the most important and classic studies on this issue is Rodger Hurley, *Poverty and Mental Retardation: A Causal Relationship* (New York: Random House, 1969). More recent studies include Richard A. Kurtz, *Social Aspects of Mental Retardation* (Lexington, Mass.: Lexington Books, 1977); Michael Maloney and Michael P. Ward, *Mental Retardation and Society* (New York: Oxford University Press, 1979); James R. Patton, James S. Payne, and Mary Beirne-Smith, *Mental Retardation*, 3rd edition (Columbus, Ohio: Merrill, 1986).
13. Rita Arditti, "The Surrogacy Business," *Social Policy* 18, no. 2 (Fall 1987):42. See also Robert Hanley, "Baby M's Best Interests May Resolve Puzzling Case," *New York Times* (February 2, 1987):B1, B3; Richard Neuhaus, "Renting Women, Buying Babies and Class Struggles," *Transaction/Society* (March/April 1988):9–29.

14. Marc Lappe, "Why Shouldn't We have a Eugenic Policy?" in *Genetics and the Law*, ed. Aubrey Milunsky and George Annas (New York: Plenum Press, 1976), 423. Part of the quotation is from M. Walzer, "In Defense of Equality," *Dissent* 20 (1973):399.
15. Edward Bellamy, *Looking Backward* (New York: Penguin Books, 1982), 40–41. (original publication in 1887).
16. C. S. Lewis, *The Abolition of Man* (New York: Macmillan, 1965), 70.
17. Martin Golding, "Ethical Issues in Biological Engineering," *UCLA Law Review* 15, no. 267 (1968):452.
18. Ibid.
19. Muller, "Should We Weaken or Strengthen Our Genetic Heritage?" 446.
20. Ibid., 445.
21. Ibid., 447.
22. Joseph Fletcher, "Indicators of Humanhood: A Tentative Profile of Man," *Hastings Center Report* 2, no. 5 (November 1972):1–4.
23. Robert Sinsheimer, "Prospects for Future Scientific Developments: Ambush or Opportunity," in *Ethical Issues in Human Genetics*, ed. Bruce Hilton et. al. (New York: Plenum Press, 1973), 350.
24. Alasdair MacIntyre, "Seven Traits for the Future," *Hastings Center Report* 9, no. 1 (February 1979):5–7.
25. Ibid., 7.
26. Roger Shinn, "Perilous Progress in Genetics," *Social Research* 4, no. 1 (Spring 1974):100.
27. Sumner Twiss, "Ethical Issues in Genetic Screening," in *Ethical, Social and Legal Dimensions of Screening for Human Genetic Disease*, ed. Daniel Bergsma (New York: National Foundation for the March of Dimes, 1974), 226.
28. Michael Lerner, "Ethics and the New Biology," in *Genetics and the Quality of Life*, 31, italics his.
29. Joshua Lederberg, "Experimental Genetics and Human Evolution," *The American Naturalist* 100, no. 915 (September/October 1966):527.
30. Joshua Lederberg, "Genetic Engineering and the Amelioration of Genetic Defect," *Science* 20, no. 24 (December 15, 1970):1310. Lederberg notes that cloning is a "speculative game" but "does serve as a metaphor to indicate that future generations will have infinitely more powerful ways than we do to deal with whatever they may regard as socially urgent issues of human nature."
31. For a discussion of cloning as a metaphor, see Willard Gaylin, "The Frankenstein Myth Becomes Reality: We Have the Awful

Knowledge to Make Exact Copies of Human Beings," *New York Times Magazine* (March 5, 1972):48.

32. Joseph Fletcher, "Ethical Aspects of Genetic Control," *New England Journal of Medicine* 285, no. 14 (September 30, 1971):779.
33. Robert Edwards and Patrick Steptoes, *A Matter of Life: The Story of a Medical Breakthrough* (New York: William Morrow, 1980), 188.
34. James Watson, "Moving Toward the Clonal Man," *Atlantic* (May 1971):52; James D. Watson, "The Future of Asexual Reproduction," *Intellectual Digest* 2, no. 2 (1971):69–74.
35. Ibid., 53.
36. See, for example, Sheryl Ruzek, *The Women's Health Movement: Feminist Alternatives to Medical Control* (New York: Praeger, 1971); Helen Marieskind, *Women in the Health System: Patients, Providers and Programs* (St. Louis: Mosby, 1980).
37. See, for example, Helen B. Holmes, Betty B. Hoskins, and Michael Gross, eds., *Birth Control and Controlling Birth: Women-Centered Perspectives* (Clifton, N.J.: Humana Press, 1980); Robert J. Apfel, *To Do No Harm: DES and the Dilemmas of Modern Medicine* (New Haven: Yale University Press, 1984).
38. See, for example, Mary Ann Warren, *Gendercide: The Implications of Sex Selection* (Totowa, N.J.: Rowman and Allenheld, 1985); Helen B.Holmes, Betty B. Hoskins, and Michael Gross, *The Custom-Made Child?: Women-Centered Perspectives* (Clifton, N.J.: Humana Press, 1981).
39. See, for example, Helen B. Holmes and Betty B. Hoskins, "Prenatal and Preconception Sex Choice Technologies: A Path to Femicide," in *Man Made Women*, ed. Gena Corea et. al. (Bloomington and Indianapolis: Indiana University Press, 1987), 15–29.
40. Marcia Guttentag and Paul F. Secord, *Too Many Women? The Sex Ratio Question* (Beverly Hills, Calif.: Sage Publications, 1983), 19.
41. See, for example, Lucille Forer, *The Birth Order Factor* (New York: David McKay, 1976); Walter Toman, *Family Constellation*, 3rd edition (New York: Springer, 1976); Robyn Roland, "Reproductive Technologies: The Final Solution to the Woman Question," in *Test-Tube Women: What Future for Motherhood?* ed. Rita Arditti, Renate Duelli Klein, and Shelley Minden (Boston: Pandora Press, 1984), 356–369.
42. John Fletcher, "Ethics and Amniocentesis for Fetal Sex Identification," *Hastings Center Report* 10, no. 1 (February 1980):15. After

initially opposing the practice, Fletcher went on to accept it on the grounds of *Roe v. Wade*.

43. Gertrude Lenzer, "Gender Ethics," *Hastings Center Report* 10, no. 1 (February 1980):18.
44. Tabitha Powledge, "Unnatural Selection: On Choosing Children's Sex," in *The Custom-Made Child?* 196–197.

CHAPTER 5. THE PROBLEM OF ALIENATION

1. In this chapter the term *alienation* will be used throughout, although it is recognized that it has many meanings and is the subject of considerable controversy. This is true even in the field of sociology, where it is a central concept. Alienation is sometimes confounded with the term *anomie,* meaning normlessness, which is a related but not identical term. In the field of bioethics, the term most typically used is *dehumanization,* deriving from philosophy. The basic meaning intended in this chapter is that of social distance or estrangement, from self, others, and nature. See, for example, Robert Nisbet, *The Sociological Tradition* (New York: Basic Books, 1966), Chapter 7.
2. Karl Marx, "Alienated Labor," trans. Eric Josephson and Mary Josephson, in *Man Alone: Alienation in Modern Society*, ed. Eric Josephson and Mary Josephson (New York: Dell, 1962), 97. For a slightly different translation, see T. B. Bottomore, ed. and trans., *Karl Marx: Selected Writings in Sociology and Social Philosophy* (New York: McGraw-Hill, 1956), 169. See also John Torrance, *Estrangement, Alienation and Exploitation: A Sociological Approach to Historical Materialism* (New York: Columbia University Press, 1977).
3. Marx, "Alienated Labor," 95.
4. Georg Simmel, "The Metropolis and Mental Life," in *Classic Essays on the Culture of Cities*, ed. Richard Sennett (Englewood Cliffs, N.J.: Prentice Hall, 1969), 49.
5. David Karp, *Being Urban* (Lexington, Mass.: Heath, 1977), 13; Ferdinand Toennies, *Fundamental Concepts of Sociology*, trans. Charles P. Loomis (New York: American Book, 1940).
6. Louis Wirth, "Urbanism as a Way of Life," in *Classic Essays on the Culture of Cities*, 153.
7. Erich Fromm, "Alienation Under Capitalism," in Josephson and Josephson, *Man Alone*, 59.
8. Cited in Josephson and Josephson, *Man Alone*, 11.

9. Stanley Joel Reiser, *Medicine and the Reign of Technology* (New York: Cambridge University Press, 1978), 229–230.
10. See, for example, Minako Maykovich, *Medical Sociology* (Sherman Oaks, Calif.: Alfred Publishing 1980), 16.
11. Samuel Osherson and Lorna Amara Singham, "The Machine Metaphor in Medicine," in *Social Contexts of Health, Illness and Patient Care*, ed. Elliot Mishler et. al. (Cambridge: Cambridge University Press, 1981), 222–223. See also Edmund D. Pellegrino and David C. Thomasma, *A Philosophical Basis of Medical Practice* (New York: Oxford University Press, 1981), 99.
12. Rene Dubos, *Mirage of Health* (New York: Harper, 1959), 152–153.
13. Although the widespread perception is that germ theory was largely responsible for these advances, there is a strong argument that public health measures, not immunizations, were responsible for most of the successes. See, for example, Thomas McKeown, *The Modern Rise of Population* (New York: Academic Press, 1976); John B. McKinlay and Sonja M. McKinlay, "Medical Measures and the Decline of Mortality," in *The Sociology of Health and Illness: Critical Perspectives*, 2nd edition, ed. Peter Conrad and Rochelle Kern (New York: St. Martin's Press, 1986), 10–23.
14. Andrew C. Twaddle and Richard M. Hessler, *A Sociology of Health* (St. Louis: Mosby, 1977), 11.
15. Richard W. Wertz and Dorothy C. Wertz, *Lying-In: A History of Childbirth in America* (New York: Free Press, 1977), 173. See also Barbara Ehrenreich and Deirdre English, *For Her Own Good: 150 Years of the Experts' Advice to Women* (Garden City, N.Y.: Doubleday, 1978); Pamela Eakins, ed., *The American Way of Birth* (Philadelphia: Temple University Press, 1986); Ann Oakley, *The Captured Womb: A History of the Medical Care of Pregnant Women* (New York: Basil Blackwell, 1984).
16. Roger Shinn, "Foetal Diagnosis and Selective Abortion: An Ethical Exploration," in *Genetics and the Quality of Life*, ed. Charles Birch and Paul Albrecht (New York: Plenum Press, 1975), 79.
17. For the most classic works on this topic, see Elizabeth Kubler Ross, *On Death and Dying* (New York: Macmillan, 1969) and Elizabeth Kubler Ross, *Questions and Answers on Death and Dying* (New York: Macmillan, 1974). See also Sandol Stoddard, *The Hospice Movement: A Better Way of Caring for the Dying* (New York: Vintage, 1978).
18. Gary Albrecht, *The Sociology of Physical Disability and Re-*

habilitation (Pittsburgh: University of Pittsburgh Press, 1976), 261.

19. Erving Goffman, *Asylums: Essays on the Social Situation of Mental Patients and Other Inmates* (Garden City, N.Y.: Anchor Books, 1961), 341–342.
20. One of the most famous and scathing of these critiques is Ivan Illich's *Medical Nemesis: The Expropriation of Health* (New York: Pantheon, 1976). A similarly strong critique of medicine's treatment of women is Gena Corea's *The Hidden Malpractice: How American Medicine Treats Women as Patients and Professionals* (New York: Morrow, 1977). For a more recent view, see also Eric J. Cassell, *The Healer's Art* (Cambridge, Mass.: M.I.T. Press, 1985).
21. See Chapter 6 for a discussion of hospital ethics committees.
22. Richard Titmuss, *The Gift Relationship: From Human Blood to Social Policy* (New York: Pantheon, 1971).
23. Ibid., 245–246.
24. Ibid., 158.
25. Ibid., 198–199, italics mine.
26. For a brief discussion of the legal question of "selling" a child, see Sue A. Meinke, "Surrogate Mothers: Ethical and Legal Issues," National Reference Center for Bioethics Literature, Kennedy Institute of Ethics, Georgetown University (August 1988).
27. Joseph Sullivan, "Brief by Feminists Opposes Surrogate Parenthood," *New York Times* (July 31, 1987):3.
28. "Test-Tube Babies: More Cases, More Marketing, More Worries," *Hastings Center Report* 15, no. 5 (October 1985):3. See also Committee to Consider The Social, Ethical and Legal Issues Arising from In Vitro Fertilization, Chairman Louis Waller, "Report on the Disposition of Embryos Produced by In Vitro Fertilization" (Melbourne, 1984).
29. Titmuss, *The Gift Relationship*, 158.
30. While the collection of *whole* blood has become increasingly voluntary, the collection of plasma is still overwhelming commercialized. In the plasmapherisis program, the red blood cells are returned to the donors, thus donors can give repeatedly. This is frequently done by low-income donors. See Piet Hagen, *Blood: Gift or Merchandise; Towards an International Blood Policy* (New York: Alan R. Liss, 1982).
31. Cited in June Goodfield, "Humanity in Science: A Perspective and a Plea," *Science* 198, no. 4317 (November 11, 1977):580.

32. *The Baby Makers*, McGraw-Hill Films (1980).
33. Paul Ramsey, "Shall We Reproduce? Rejoinders and Future Forecast," Part II, *Journal of the American Medical Association* 220, no. 11 (June 12, 1972):1485.
34. Patricia Spallone, *Beyond Conception: The New Politics of Reproduction* (Granby, Mass.: Bergin and Garvey Publishers, 1989), 96.
35. Ibid., 97–98.
36. Clifford Grobstein, "The Moral Uses of Spare Embryos," *Hastings Center Report* 12, no. 3 (June 1982):5.
37. Cited in Peter Williams and Gordon Stevens, "What Now for Test-Tube Babies?" *New Scientist* 93, no. 1291 (February 4, 1982):314.
38. *The Baby Makers*.
39. Amatai Etzioni, *Genetic Fix* (New York: Macmillan, 1975).
40. Jacques Ellul, *The Technological Society*, trans. John Wilkinson (New York: Vintage, 1964), 79.
41. Joseph Fletcher, "Ethical Aspects of Genetic Controls," *New England Journal of Medicine* 285, no. 14 (September 30, 1971):780–781.
42. Leon Kass, "The New Biology: What Price Relieving Man's Estate?" in *Bioethics*, ed. Thomas A. Shannon (New York: Paulist Press, 1976), 309–310, italics his.
43. Ibid., 312.
44. See, for example, Barbara Katz Rothman, "The Social Construction of Birth," in *The American Way of Birth*, 104–118; David Sudnow, *Passing On: The Social Organization of Dying* (Englewood Cliffs, N.J.: Prentice Hall, 1967).
45. Oakley, *The Captured Womb*, 276.
46. Ruth Hubbard, "Personal Courage Is Not Enough: Some Hazards of Childbearing in the 1980s," in *Test-Tube Women: What Future for Motherhood?* ed. Rita Arditti, Renate Duelli Klein, and Shelley Minden (Boston: Pandora Press, 1984), 332.
47. Ellul, *The Technological Society*, 422.

CHAPTER 6. SCIENCE, REGULATION, AND PUBLIC POLICY

1. See James Watson, *The Double Helix* (New York: Atheneum, 1968).
2. See, for example, James Jones, *Bad Blood: The Tuskegee Syphilis Experiment* (New York: Free Press, 1981); Allan M. Brandt, "Racism and Research: The Case of the Tuskegee Syphilis Study," *Hastings Center Report* 8, no. 6 (December 1978)): 21–29.

3. Ibid.
4. See, for example, Robert G. Weisbord, *Genocide: Birth Control and the Black American* (Westport, Conn.: Greenwood Press, 1975), 36.
5. See David Rothman, "Were Tuskegee and Willowbrook 'Studies in Nature'?" *Hastings Center Report* 12, no. 2 (April 1982):5–7.
6. U.S. Congress, Office of Technology Assessment, *The Regulatory Environment for Science—a Technical Memorandum*, OTA-TM-SET-34 (Washington, D.C.: Government Printing Office, 1986), 22.
7. Robert Sinsheimer, "The Presumptions of Science," *Daedalus* (Spring 1978):23; for a more recent discussion of the "limits to inquiry" debate, see *The Regulatory Environment for Science—a Technical Memorandum*, 23–26.
8. Ibid.
9. Jeremy Rabkin, "Office for Civil Rights," in *The Politics of Regulation*, ed. James Q. Wilson (New York: Basic Books, 1980), 304.
10. Jeremy Cherfas, *Man Made Life: An Overview of the Science, Technology and Commerce of Genetic Engineering* (New York: Pantheon, 1982); see also Clifford Grobstein, "The Recombinant-DNA Debate," *Scientific American* 237, no. 1 (July 1977):22–33.
11. Maxine Singer and Dieter Soll, "Guidelines for DNA Hybrid Molecules," *Science* 181, no. 4105 (September 21, 1973):1114.
12. Cherfas, *Man Made Life*; Paul Berg et. al., "Potential Biohazards of Recombinant DNA Molecules," *Science* 185, no. 4148 (July 26, 1974):303.
13. This occurred in 1976. See R. W. Old and S. B. Primrose, *Principles of Gene Manipulation* (Berkeley: University of California Press, 1980), 115.
14. "U.S. Recombinant DNA Guidelines: Expected Change," *Nature* 296, no. 29 (April 1982):793.
15. Clifford Grobstein, "Asilomar and the Formation of Public Policy," in *The Gene-Splicing Wars: Reflections in the Recombinant DNA Controversy*, ed. Raymond A. Zilinskas and Burke K. Zimmerman (New York: Macmillan, 1986), 6.
16. Cited in George H. Kieffer, *Bioethics: A Textbook of Issues* (Reading, Mass.: Addison Wesley, 1979), 424.
17. Diana B. Dutton and John L. Hochheimer, "Institutional Biosafety Committees and Public Participation: Assessing an Experiment," *Nature* 297, no. 6 (May 1982):11.
18. Ibid.

19. Claire Nadler, "Technology and Democratic Control," in *The Gene-Splicing Wars*, 148.
20. Ronald E. Cranford and A. Edward Doudera, "The Emergence of Institutional Ethics Committees," in *Institutional Ethics Committees and Health Care Decisionmaking*, ed. Ronald E. Cranford and A. Edward Doudera (Ann Arbor, Mich.: Health Administration Press, 1984), 7. See also Karen Teel, "The Physician's Dilemma: What the Law Should Be," *Baylor Law Review* 27 (Winter 1975) 6, 8–10; Robert Veatch, "Hospital Ethics Committees: Is There a Role?" *Hastings Center Report* 7, no. 3 (June 1977):22–25; Carol Levine, "Hospital Ethics Committees: A Guarded Prognosis," *Hastings Center Report* 7, no. 3 (June 1977):25–26; Judith Randal, "Are Ethics Committees Alive and Well?" *Hastings Center Report* 13, no. 6 (December 1983):10–12.
21. Cranford and Doudera, "The Emergence of Institutional Ethics Committees," 5; see also Alan R. Fleischman and Thomas H. Murray, "Ethics Committees for Infants Doe?" *The Hastings Center Report* 13, no. 6 (December 1983):5–9; Alan R. Fleischman, "An Infant Bioethical Review Committee in an Urban Medical Center," *Hastings Center Report* 16, no. 3 (June 1986):16–18.
22. See Renee Fox and Judith Swazey, *The Courage to Fail* (Chicago: University of Chicago Press, 1978); Howard Brody, *Ethical Decisions in Medicine*, 2nd edition (Boston: Little, Brown, 1981), 224.
23. Policies currently under consideration would go far beyond the Uniform Anatomical Gift Act of 1968, which permitted persons to donate organs upon death, and would require or strongly encourage donations.
24. The Diagnostic Related Group (DRG) system is a hospital financing system in which each disease is granted a specified reimbursement, irrespective of the patient's actual condition. It is used by Medicare and some private insurance programs.
25. See, for example, Susan M. Wolf, "Ethics Committees in the Courts," *Hastings Center Report* 10, no. 3 (June 1986):12, who notes "so far the Courts of three states have grappled with the question of how to treat an ethics committee's determination; and the highest courts of three states have given three different answers. The Georgia Court ignored the committee's determination; the Minnesota Court treated Committee determinations as evidence; and the Massachusetts courts indicated that a judge has the option to treat a committee determination as evidence, but implied that this evidence is relevant to considerations other than those suggested by the Minnesota court."

26. See, for example, "AMA Judicial Chairman Sees Emergence of Hospital Ethics Panels as an Inevitability," *Federation of American Hospitals Review* 16 (November/December 1983):31.
27. John A. Robertson, "The Law of Institutional Review Boards," *UCLA Law Review* 26, no. 3 (February 1979):484.
28. Jerry Mashaw, "Thinking about Institutional Review Boards," in *Whistleblowing in Biomedical Research: Proceedings of a Workshop*, President's Commission for the Study of Ethical Problems in Medicine and Biomedical and Behavioral Research (Washington, D.C.: Government Printing Office, 1983).
29. President's Commission for the Study of Ethical Problems in Medicine and Biomedical and Behavioral Research, *Compensating for Research Injuries, Vols. I, II; Deciding to Forego Life-Sustaining Treatment; Defining Death; Implementing Human Research Regulations; Making Health Care Decisions, Vols. I, II, III; Protecting Human Subjects; Screening and Counseling for Genetic Conditions; Securing Access to Health Care, Vols. I, II, III; Splicing Life: The Social and Ethical Issues of Genetic Engineering with Human Beings; Whistle Blowing in Biomedical Research, Summing Up* (Washington, D.C.: Government Printing Office, 1982, 1983).
30. See *Splicing Life: The Social and Ethical Issues of Genetic Engineering with Human Beings*, 7.
31. Alexander Morgan Capron, "Looking Back at the President's Commission," *Hastings Center Report* 13, no. 5 (October 1983):7–12.
32. Cranford and Doudera, "The Emergence of Institutional Ethics Committees," 6.
33. U.S. Riot Commission Report, *Report of the National Advisory Commission on Civil Disorders* (Washington, D.C.: Government Printing Office, 1968).
34. "Designing Life, by the Rules," *New York Times* (August 24, 1987):A18.
35. Ibid.
36. Keith Schneider, "Germ Scientist Freed Germs in 1984 Test," *New York Times* (September 1, 1987):1.
37. "Designing Life, by the Rules."
38. Donald S. Frederikson, "The Recombinant DNA Controversy: The NIH Viewpoint," in *The Gene-Splicing Wars*, 22.
39. Richard McCormick, "Ethics Committees: Promise or Peril," *Law, Medicine and Health Care* 12, no. 4 (September 1984):153.
40. David Thomasma, "Hospital Ethics Committees: Laying the

Groundwork," in *Bioethics Reporter: Ethical and Legal Issues in Medicine, Health Care Administration and Human Experimentation*, ed. James Childress et. al. (Frederick, Md.: University Publications of American, 1984), 706.

CHAPTER 7. CITIZEN PARTICIPATION AND PUBLIC POLICY

1. See, for example, Peter Bachrach and Morton Baratz, "The Two Faces of Power," *American Political Science Review* 57 (December 1962):947–952; C. Wright Mills, *The Power Elite* (New York: Oxford University Press, 1959).
2. Peter Bachrach and Morton Baratz, *Power and Poverty: Theory and Practice* (New York: Oxford University Press, 1970), 44–45.
3. Sheldon Krimsky, *Genetic Alchemy: The Social History of the Recombinant DNA Controversy* (Cambridge, Mass.: M.I.T. Press, 1982), 294.
4. Cited in *The Recombinant DNA Debate*, ed. David A. Jackson and Stephen Stitch (Englewood Cliffs, N.J.: Prentice Hall, 1979), 103.
5. The Cambridge protest was actually the second major event that occurred. The first took place at the University of Michigan in 1974. This protest was primarily directed by university faculty and was thus more of an "in-house" and institutionally directed protest than an externally generated action. See Krimsky, *Genetic Alchemy*.
6. Nicholas Wade, "Gene-Splicing: Cambridge Citizens OK Research but Want More Safety," *Science* 195, no. 4275 (January 21, 1977):268.
7. Ibid.
8. Ibid.
9. David Clem, "Regulation at Cambridge," in *Recombinant DNA: Science, Ethics, and Politics*, ed. John Richards (New York: Academic Press, 1978), 244.
10. Wade, "Gene-Splicing: Cambridge Citizens OK Research." See also Krimsky, *Genetic Alchemy*, for a personal account of the CERB proceedings. He suggests there were more than 100 hours of testimony.
11. Clem, "Regulation at Cambridge," 246.
12. Wade, "Gene-Splicing: Cambridge Citizens OK Research," 269.
13. Clem, "Regulation at Cambridge," 242.

14. U.S. Congress, Office of Technology Assessment, *The Regulatory Environment for Science—a Technical Memorandum*, OTA-TM-SET-34 (Washington, D.C.: Government Printing Office, 1986), 108; Nicholas Wade, "Gene-Splicing: At Grass-Roots Level a Hundred Flowers Bloom," *Science* 195, no. 4278 (February 11, 1977):558–560.
15. Cited in Krimsky, *Genetic Alchemy*, 301.
16. U.S. Congress, Office of Technology Assessment, *The Regulatory Environment for Science—a Technical Memorandum*, 100.
17. Ibid., 101.
18. Ibid.
19. Andrew Burness, "Who Lives, Who Dies, Who Pays?" *Foundation News* 27, no. 5 (September/October 1986):36.
20. The following states either have existing projects or are contemplating projects: Arizona, California, Colorado, Florida, Hawaii, Idaho, Illinois, Iowa, Kansas, Maine, Minnesota, Nebraska, New Jersey, New Mexico, Vermont, Washington, Wisconsin.
21. Burness, "Who Lives, Who Dies, Who Pays?" 36.
22. Oregon Health Decisions, *Reporter* (October 1987):2; Ralph Crenshaw et. al., "Oregon Health Decisions," *Journal of the American Medical Association* 254, no. 22 (December 13, 1985):3213–3216.
23. Bruce Jennings, "Community Bioethics: Notes on a New Movement," *Federation Review* 9, no. 5 (September/October 1986):19.
24. Brian L. Hines, *Oregon and American Health Decisions: A Guide for Community Action on Bioethical Issues* (July 1985):6.
25. Oregon Health Decisions, *Reporter* (April 1988):1. For structures in other states, see, for example: Leonard M. Fleck, *Just Caring: Justice, Health Care, and the Good Society, an Ethics Forum* (Goshen, Ind.: Goshen General Hospital, no date); Idaho Health Systems, *No Easy Choices . . :Ethical Dilemmas in Health Care Resource Allocation*, results of the Citizens Health Care Parliament (Boise, Idaho: Idaho Health Systems Agency, 1987); Vermont Health Policy Council, *Taking Steps: Ethical Decisions for Living and Dying*, final report (Waterbury, VT.: Vermont Health Policy Council, December, 1987). Susan Dade et al., *Washington Health Choices: Involving the Public in Health Care Choices* (Seattle, Wash.: Puget Sound Health Systems Resources, no date).
26. Hines, *Oregon and American Health Decisions*, 6.
27. Ibid.

28. Oregon Health Decisions, *Reporter* (May 1986):6.
29. Hines, *Oregon and American Health Decisions*, 9–10.
30. Bruce Jennings, "A Grassroots Movement in Bioethics," *Hastings Center Report*, Special Supplement 18, no. 3 (June/July 1988):3.
31. Ibid., 2.
32. Hines, *Oregon and American Health Decisions*, 9.
33. Burness, "Who Lives, Who Dies, Who Pays?" 36.
34. Jennings, "A Grassroots Movement in Bioethics," 3.
35. Ibid., 1.
36. Judith P. Swazey and Kathleen Lawrence, *Many Shades of Gray: Health Care Needs and Services for Maine's Elderly*, Maine Health Care Decisions, final report (October 1987).
37. Oregon Health Decisions, *Reporter* (May 1986):15.
38. Sheldon Krimsky, "The Corporate Capture of Academic Science and Its Social Costs," in *Genetics and the Law, III*, ed. Aubrey Milunsky and George Annas (New York: Plenum Press, 1985), 46.
39. Amatai Etzioni and Clyde Nunn, "The Public Appreciation of Science in Contemporary America," *Daedalus* 103 (1974):193.
40. Ibid., 194.
41. John Walsh, "Public Attitude Toward Science Is Yes, But—," *Science* 215, no. 4530 (January 15, 1982):270.
42. U.S. Congress, Office of Technology Assessment, *The Regulatory Environment for Science—a Technical Memorandum*, 130.
43. Ibid., 132.
44. Ibid., 133.
45. Ibid., 132.
46. Jack DeSario and Stuart Langton, "Citizen Participation and Technocracy," in *Citizen Participation in Public Decision Making*, ed. Jack DeSario and Stuart Langton (New York: Greenwood Press, 1987), 11.
47. Ibid., 12.
48. Etzioni and Nunn, "The Public Appreciation of Science," 195.
49. U.S. Congress, Office of Technology Assessment, *New Developments in Biotechnology—Background Paper: Public Perceptions of Biotechnology*, OTA-BP-BA-45 (Washington, D.C.: Government Printing Office, May 1987), 2.
50. Etzioni and Nunn, "The Public Appreciation of Science," 196.
51. Walsh, "Public Attitude Toward Science Is Yes, But—," 270.
52. Ibid., 272.
53. U.S. Congress, Office of Technology Assessment, *The Regulatory Environment for Science—a Technical Memorandum*, 131.

54. Ibid., 195.
55. Robert Morrison, "Commentary on 'The Boundaries of Scientific Freedom,'" *Newsletter on Science, Technology and Human Values* (June 1977):24, cited in U.S. Congress, Office of Technology Assessment, *The Regulatory Environment for Science—a Technical Memorandum*, 25.
56. James D. Watson and John Tooze, *The DNA Story: A Documentary History of Gene Cloning* (San Francisco: Freeman, 1981), ix.
57. Gerald Holton, "Epilogue to the Issues, Limits of Scientific Inquiry," *Daedalus* (Spring 1978):232.
58. Dorothy Nelkin, "Threats and Promises: Negotiating the Control of Research," *Daedalus* (Spring 1978):203.
59. Daniel Callahan, "The Involvement of the Public," in *Research with Recombinant DNA: An Academy Forum* (Washington, D.C.: National Academy of Sciences, 1977), 35, italics mine.
60. Nelkin, "Threats and Promises," 192.
61. Clement Bezold, "Beyond Technocracy: Anticipatory Democracy in Government and the Marketplace" in *Citizen Participation in Public Decision Making*, 66.

CHAPTER 8. BALANCING THE SOCIAL AND GENETIC AGENDAS

1. Dennis Wrong, *Population and Society*, 3rd edition (New York: Random House, 1967), 31.
2. Children's Defense Fund, *The Health of America's Children: Maternal and Child Health Data Book* (Washington D.C., 1989), x, emphasis mine.
3. Ibid.
4. "The Latest Word," *Hastings Center Report* 18, no. 3 (June/July 1988):45.
5. Children's Defense Fund, *The Health of America's Children*, 48. See also Ruth Sidel, *Women and Children Last: The Plight of Poor Women in Affluent America* (New York: Viking, 1986); Hilda Scott, *Working Your Way to the Bottom: The Feminization of Poverty* (Boston: Pandora Press, 1984).
6. The Children's Defense Fund, *The Health of America's Children*, 36, 50, 51.
7. For example, according to the Children's Defense Fund, "Death rates among black and white infants resulting from congenital anomalies (largely nonpreventable birth defects) were virtually

identical. . . . However, for many causes considered preventable through adequate maternity and infant health care, black infants were substantially more likely to die than white infants. Black infants were nearly four times more likely to die as a result of prematurity and low birthweight . . . two and a half times more likely to die of pneumonia or influenza, and two times more likely to die as a result of newborn and maternal complications, newborn infections, and accidents." Children's Defense Fund, *The Health of America's Children*, 9.

8. See David Stoesz and Howard Karger, *American Social Welfare Policy: A Structural Approach* (New York: Longman, 1990), 198. Washington D.C. is compared with states, not cities, and it ranks worse than all other states.
9. Martin Tolchin, "Richest Got Richer and Poorest Poorer in 1979–1987," *New York Times* (March 23, 1989):A1. See also Kevin Phillips, *The Politics of Rich and Poor: Wealth and The American Electorate in the Reagan Aftermath* (New York: Random House, 1990).
10. Ibid.
11. Leon Jaroff, "The Gene Hunt," *Time* (March 20, 1989):67.
12. Ibid., 63.
13. Ibid.
14. Ibid.
15. Figures for first three statistics cited from Children's Defense Fund, *A Vision for America's Future* (Washington, D.C., 1989), 28, 38, 113.
16. Children's Defense Fund, *The Health of America's Children*, xiii.
17. Iris Shannon, "Environmental Health and Children," *The Nation's Health* (July 1989):2.
18. Jaroff, "The Gene Hunt," 64.
19. Ibid., 67.
20. John B. McKinlay, "A Case for Refocusing Upstream: The Political Economy of Illness," in *The Sociology of Health and Illness: Critical Perspectives*, 2nd edition (New York: St. Martin's Press, 1986), 484–485, emphasis his.
21. David E. Reiser and David H. Rosen, *Medicine as a Human Experience* (Baltimore: University Park Press, 1984), x.
22. Ibid.
23. Rosalind Petchesky, "Foetal Images: The Power of Visual Culture in the Politics of Reproduction," in *Reproductive Technologies: Gender, Motherhood, and Medicine*, ed. Michelle Stanworth (Minneapolis: University of Minnesota Press, 1987), 61–62.

24. David Kevles, *In the Name of Eugenics: Genetics and the Uses of Human Heredity* (New York: Knopf, 1985), ix.
25. Nancy S. Wexler, "Will the Circle Be Unbroken? Sterilizing the Genetically Impaired," in *Genetics and the Law, II*, ed. Aubrey Milunsky and George J. Annas (New York: Plenum Press, 1980), 315.
26. Michelle Stanworth, "The Deconstruction of Motherhood," in *Reproductive Technologies*, 28.
27. Linda Gordon, *Woman's Body, Woman's Right: A Social History of Birth Control in America* (New York: Penguin, 1976), 130.
28. Harold L. Wilensky and Charles Lebeaux, *Industrial Society and Social Welfare* (New York: Russell Sage, 1958), 138–140.
29. Robert Sinsheimer, "An Evolutionary Perspective for Genetic Engineering," *New Scientist* 73, no. 1035 (January 20, 1977):152.
30. Robert Bellah, *Habits of the Heart: Individualism and Commitment in American Life* (New York: Harper and Row, 1985), 284.
31. For a interesting history and analysis of this community, see Benjamin Zablocki, *The Joyful Community: An Account of the Bruderhof, A Communal Movement Now in Its Third Generation* (Baltimore: Penguin, 1971).
32. Diana B. Dutton and John L. Hochheimer, "Institutional Biosafety Committees and Public Participation: Assessing an Experiment," *Nature* 297, no. 5861 (May 6, 1982):15.

Index